REVIEW OF THE RESEARCH PROGRAM OF THE **FreedomCAR** AND **Fuel Partnership**

SECOND REPORT

Committee on Review of the FreedomCAR and Fuel Research Program, Phase 2

Board on Energy and Environmental Systems

Division on Engineering and Physical Sciences

NATIONAL RESEARCH COUNCIL
OF THE NATIONAL ACADEMIES

THE NATIONAL ACADEMIES PRESS
Washington, D.C.
www.nap.edu

THE NATIONAL ACADEMIES PRESS 500 Fifth St., N.W. Washington, DC 20001

This report and the study on which it is based were supported by Contract No. DE-AT-01-06EE11206, TO#18, Subtask 2. Any opinions, findings, conclusions, or recommendations expressed in this publication are those of the author(s) and do not necessarily reflect the views of the organizations or agencies that provided support for the project.

International Standard Book Number-13: 978-0-309-11634-3
International Standard Book Number-10: 0-309-11634-1

Available in limited supply from
Board on Energy and Environmental Systems
National Research Council
500 Fifth Street, N.W.
Keck W934
Washington, DC 20001
202-334-3344

Additional copies are available for sale from
The National Academies Press
2101 Constitution Avenue, N.W.
Lockbox 285
Washington, DC 20055
800-624-6242 or 202-334-3313 (in the Washington metropolitan area)
http://www.nap.edu

Printed in the United States of America

THE NATIONAL ACADEMIES

Advisers to the Nation on Science, Engineering, and Medicine

The **National Academy of Sciences** is a private, nonprofit, self-perpetuating society of distinguished scholars engaged in scientific and engineering research, dedicated to the furtherance of science and technology and to their use for the general welfare. Upon the authority of the charter granted to it by the Congress in 1863, the Academy has a mandate that requires it to advise the federal government on scientific and technical matters. Dr. Ralph J. Cicerone is president of the National Academy of Sciences.

The **National Academy of Engineering** was established in 1964, under the charter of the National Academy of Sciences, as a parallel organization of outstanding engineers. It is autonomous in its administration and in the selection of its members, sharing with the National Academy of Sciences the responsibility for advising the federal government. The National Academy of Engineering also sponsors engineering programs aimed at meeting national needs, encourages education and research, and recognizes the superior achievements of engineers. Dr. Charles M. Vest is president of the National Academy of Engineering.

The **Institute of Medicine** was established in 1970 by the National Academy of Sciences to secure the services of eminent members of appropriate professions in the examination of policy matters pertaining to the health of the public. The Institute acts under the responsibility given to the National Academy of Sciences by its congressional charter to be an adviser to the federal government and, upon its own initiative, to identify issues of medical care, research, and education. Dr. Harvey V. Fineberg is president of the Institute of Medicine.

The **National Research Council** was organized by the National Academy of Sciences in 1916 to associate the broad community of science and technology with the Academy's purposes of furthering knowledge and advising the federal government. Functioning in accordance with general policies determined by the Academy, the Council has become the principal operating agency of both the National Academy of Sciences and the National Academy of Engineering in providing services to the government, the public, and the scientific and engineering communities. The Council is administered jointly by both Academies and the Institute of Medicine. Dr. Ralph J. Cicerone and Dr. Charles M. Vest are chair and vice chair, respectively, of the National Research Council.

www.national-academies.org

COMMITTEE ON REVIEW OF THE FREEDOMCAR AND FUEL RESEARCH PROGRAM, PHASE 2

CRAIG MARKS, *Chair*, NAE,[1] Altarum, Bloomfield Hills, Michigan
PETER BEARDMORE, NAE, Ford Motor Company (retired), West Bloomfield, Michigan
DAVID L. BODDE, Clemson University, South Carolina
GLENN A. EISMAN, Rensselaer Polytechnic Institute, Troy, New York
W. ROBERT EPPERLY, Consultant, Mountain View, California
DAVID E. FOSTER, University of Wisconsin, Madison
JOHN B. HEYWOOD, NAE, Massachusetts Institute of Technology, Cambridge
HAROLD H. KUNG, Northwestern University, Evanston, Illinois
JAMES J. MacKENZIE, World Resources Institute (retired), Washington, D.C.
CHRISTOPHER L. MAGEE, NAE, Massachusetts Institute of Technology, Cambridge
ROBERT J. NOWAK, Defense Advanced Research Projects Agency (retired), Rehoboth Beach, Delaware
MICHAEL P. RAMAGE, NAE, ExxonMobil Research and Engineering Company (retired), Moorestown, New Jersey
VERNON P. ROAN, University of Florida (professor emeritus), Gainesville
BERNARD ROBERTSON, NAE, DaimlerChrysler Corporation (retired), Bloomfield Hills, Michigan
R. RHOADS STEPHENSON, NASA Jet Propulsion Laboratory (retired), Consultant, La Cañada, California
KATHLEEN C. TAYLOR, NAE, General Motors Corporation (retired), Falmouth, Massachusetts
GIRI VENKATARAMANAN, University of Wisconsin, Madison
BRIJESH VYAS, Alcatel-Lucent, Murray Hill, New Jersey

Subgroup on Systems Analysis and Simulation

JOHN B. HEYWOOD, *Lead*
PETER BEARDMORE
DAVID L. BODDE
DAVID E. FOSTER
HAROLD H. KUNG
CHRISTOPHER L. MAGEE
BERNARD ROBERTSON

[1]National Academy of Engineering.

Subgroup on Advanced Combustion Engines, Emissions Control, and Hydrocarbon Fuels

DAVID E. FOSTER, *Lead*
JOHN B. HEYWOOD
HAROLD H. KUNG
MICHAEL P. RAMAGE
BERNARD ROBERTSON
KATHLEEN C. TAYLOR

Subgroup on Electrochemical Energy Storage

BRIJESH VYAS, *Lead*
CHRISTOPHER L. MAGEE
ROBERT J. NOWAK
GIRI VENKATARAMANAN

Subgroup on Fuel Cells

GLENN A. EISMAN, *Lead*
ROBERT J. NOWAK
VERNON P. ROAN
KATHLEEN C. TAYLOR
BRIJESH VYAS

Subgroup on Electric Propulsion, Electrical Systems, and Power Electronics

GIRI VENKATARAMANAN, *Lead*
BERNARD ROBERTSON
BRIJESH VYAS

Subgroup on Hydrogen Production, Delivery, and Off-board Storage

W. ROBERT EPPERLY, *Lead*
DAVID L. BODDE
GLENN A. EISMAN
HAROLD H. KUNG
JAMES J. MacKENZIE
MICHAEL P. RAMAGE

Subgroup on Onboard Hydrogen Storage

KATHLEEN C. TAYLOR, *Lead*
DAVID L. BODDE
CHRISTOPHER L. MAGEE
ROBERT J. NOWAK
VERNON P. ROAN
R. RHOADS STEPHENSON

Subgroup on Safety

R. RHOADS STEPHENSON, *Lead*
DAVID L. BODDE
W. ROBERT EPPERLY
HAROLD H. KUNG
CHRISTOPHER L. MAGEE

Subgroup on Materials

PETER BEARDMORE, *Lead*
GLENN A. EISMAN
CHRISTOPHER L. MAGEE
KATHLEEN C. TAYLOR

Project Staff

JAMES ZUCCHETTO, Director, Board on Energy and Environmental Systems (BEES)
PANOLA GOLSON, Program Associate (BEES) (until March 2007)
KATHERINE BITTNER, Senior Program Assistant
LaNITA JONES, Program Associate

BOARD ON ENERGY AND ENVIRONMENTAL SYSTEMS

DOUGLAS M. CHAPIN, *Chair*, NAE,[1] MPR Associates, Inc., Alexandria, Virginia
ROBERT W. FRI, *Vice-Chair*, Resources for the Future (senior fellow emeritus), Washington, D.C.
RAKESH AGRAWAL, NAE, Purdue University, West Lafayette, Indiana
ALLEN J. BARD, NAS,[2] University of Texas, Austin
ANDREW BROWN, JR., NAE, Delphi Corporation, Troy, Michigan
MARILYN BROWN, Georgia Institute of Technology, Atlanta
PHILIP R. CLARK, NAE, GPU Nuclear Corporation (retired), Boonton, New Jersey (term ended July 31, 2007)
MICHAEL L. CORRADINI, NAE, University of Wisconsin, Madison
PAUL DeCOTIS, New York State Energy Research and Development Authority, Albany
E. LINN DRAPER, JR., NAE, American Electric Power, Inc. (emeritus), Austin, Texas
CHARLES H. GOODMAN, Southern Company (retired), Birmingham, Alabama
DAVID G. HAWKINS, Natural Resources Defense Council, Washington, D.C.
JAMES MARKOWSKY, NAE, Consultant, North Falmouth, Massachusetts
DAVID K. OWENS, Edison Electric Institute, Washington, D.C.
WILLIAM F. POWERS, NAE, Ford Motor Company (retired), Ann Arbor, Michigan
TONY PROPHET, Carrier Corporation, Farmington, Connecticut (term ended July 31, 2007)
MICHAEL P. RAMAGE, NAE, ExxonMobil Research & Engineering Company (retired), Moorestown, New Jersey
MAXINE SAVITZ, NAE, Honeywell, Inc. (retired), Los Angeles, California
PHILIP R. SHARP, Resources for the Future, Washington, D.C. (term ended July 31, 2007)
SCOTT W. TINKER, University of Texas, Austin

Staff

JAMES ZUCCHETTO, Director
KATHERINE BITTNER, Senior Program Assistant
MATT BOWEN, Senior Program Associate (until November 2007)
DUNCAN BROWN, Senior Program Officer
JENNIFER BUTLER, Financial Assistant (until December 2007)

---

[1]National Academy of Engineering.
[2]National Academy of Sciences.

DANA CAINES, Financial Associate
ALAN CRANE, Senior Program Officer
PANOLA GOLSON, Program Associate (until May 2007)
JOHN HOLMES, Senior Program Officer
LaNITA JONES, Program Associate
MARTIN OFFUTT, Senior Program Officer (until April 2007)
MADELINE WOODRUFF, Senior Program Officer

Preface

As outlined in the Partnership Plan, the FreedomCAR and Fuel Research Partnership is a major long-term research effort whose ultimate goal is to enable the full spectrum of light-duty passenger vehicle classes to operate completely free of petroleum and free of harmful emissions while sustaining the driving public's freedom of mobility and freedom of vehicle choice. This research is directed and supported by a collaboration among the U.S. government, in particular the Department of Energy (DOE); the U.S. Council for Automotive Research (USCAR), whose members are Chrysler LLC, the Ford Motor Company, and General Motors Corporation; and five key energy companies: BP America, Chevron Corporation, ConocoPhillips, ExxonMobil Corporation, and Shell Hydrogen (U.S.). During the past 4 years, this Partnership has established a roadmap with a detailed set of research goals and milestones and has funded projects to enable progress toward its very ambitious ultimate goal, which is of critical strategic importance to the United States and to each of the companies involved.

This report is the result of the second biennial review of the progress of this program by the Committee on Review of the FreedomCAR and Fuel Research Program, Phase 2, chartered by the National Research Council (NRC). It assesses the structure and management of the program, as well as the nature, adequacy, and progress of the research activities being conducted. Critique and recommendations are provided for each of the areas assessed with the intent of enhancing the progress of this very important program.

Craig Marks
Chairman

Acknowledgments

The committee wishes to thank members of the FreedomCAR and Fuel Partnership, all of whom contributed a significant amount of their time and effort to this National Research Council (NRC) study by giving presentations at meetings, responding to requests for information, or providing valuable input. Finally, the chair wishes to recognize the committee members and the staff of the Board on Energy and Environmental Systems for their hard work in organizing and planning committee meetings and their individual efforts in gathering information and writing sections of the report.

This report has been reviewed in draft form by individuals chosen for their diverse perspectives and technical expertise, in accordance with procedures approved by the NRC's Report Review Committee. The purpose of this independent review is to provide candid and critical comments that will assist the institution in making its published report as sound as possible and to ensure that the report meets institutional standards for objectivity, evidence, and responsiveness to the study charge. The review comments and draft manuscript remain confidential to protect the integrity of the deliberative process. We wish to thank the following individuals for their review of this report:

William Agnew, NAE, General Motors Research Laboratories, retired,
Andrew Brown, NAE, Delphi Corporation,
Tom Cackette, California Air Resources Board,
Charles H. Goodman, Southern Company Services, Inc., retired,
Julius Harwood, NAE, Ford Motor Company, retired,
Fritz R. Kalhammer, Electric Power Research Institute, retired,
John Kassakian, NAE, Massachusetts Institute of Technology,

Gene Nemanich, Consultant, and ChevronTexaco Ventures, retired,
Dan Sperling, University of California, Davis,
Rodney Tabaczynski, NAE, RJ Technologies, LLC, and
John J. Wise, NAE, Mobil Research and Development Corporation, retired.

Although the reviewers listed above have provided many constructive comments and suggestions, they were not asked to endorse the conclusions or recommendations, nor did they see the final draft of the report before its release. The review of this report was overseen by Elisabeth M. Drake, NAE, Massachusetts Institute of Technology. Appointed by the National Research Council, she was responsible for making certain that an independent examination of this report was carried out in accordance with institutional procedures and that all review comments were carefully considered. Responsibility for the final content of this report rests entirely with the authoring committee and the institution.

Contents

APPENDIXES

Tables and Figures

TABLES

FIGURES

Summary

THE FREEDOMCAR AND FUEL PARTNERSHIP

This is the report of the Committee on Review of the FreedomCAR and Fuel Research Program, Phase 2, chartered by the National Research Council (NRC). The Phase 1 review of the Partnership was published by the NRC in 2005. The FreedomCAR and Fuel Partnership is a collaboration among the U.S. government, in particular the Department of Energy (DOE); the U.S. Council for Automotive Research (USCAR), whose members are Chrysler LLC, the Ford Motor Company, and General Motors Corporation; and five key energy companies: BP America, Chevron Corporation, ConocoPhillips, ExxonMobil Corporation, and Shell Hydrogen (U.S.). The program supports the very wide variety of research activities needed to enable a transition pathway for automotive transportation. The pathway starts with internal combustion engines (ICEs) more efficient than today's, proceeds through the increasing use of a variety of ICE hybrid electric vehicles, and then, by 2015, arrives at the point where the private sector can make a decision, based on information generated by the Partnership, about the commercialization of fuel-cell-powered vehicles fueled by economically competitive hydrogen produced from a variety of energy sources. Research goals have been established that, if achieved, promise to overcome the many high-risk barriers to achieving this vision.

A major strength of the FreedomCAR and Fuel Partnership is that the research it sponsors is determined by joint industry/government teams. This collaborative structure allows identifying both the long-range, precompetitive research needs, as envisioned by the automotive and energy companies, and the nation's societal needs related to automotive vehicles and fuels, as articulated by government, setting appropriate goals, and selecting the best way of achieving them. Such a collaboration is intended to speed the market deployment of radically new systems on a large scale.

The FreedomCAR and Fuel Partnership started with a presidential commitment to request $1.7 billion over 5 years (FY04 through FY08), with appropriations thus far of about $243 million, $307 million, and $339 million in FY04, FY05, and FY06, respectively. The FY07 Continuing Resolution resulted in funding of about $401 million. The FY08 Presidential Budget Request is for about $436 million. These funds are used to support basic research, applied research, development, and technology validation and deployment in the following areas:

- ICEs using a variety of fuels,
- Fuel cell power systems,
- Hydrogen storage systems,
- Electrochemical energy storage,
- Electric propulsion systems,
- Hydrogen production and delivery systems, and
- Materials for lightweight vehicles.

Specific research goals to be met in 2010 and 2015 have been established in each of these areas.

There are 11 technical teams consisting of scientists and engineers from the USCAR member companies, energy partner companies, and national laboratories, as well as DOE managers of technology development:

- Advanced combustion and emission control,
- Fuel cells,
- Onboard hydrogen storage,
- Electrochemical energy storage,
- Electrical and electronics,
- Materials,
- Hydrogen production,
- Hydrogen delivery,
- Fuel pathway integration,
- Codes and standards (C&S), and
- Vehicle systems analysis.

Program oversight is provided by an Executive Steering Group consisting of the DOE assistant secretary for energy efficiency and renewable energy (EERE) and a vice-presidential-level executive from each of the Partnership companies. Within EERE, the DOE efforts are divided between the FreedomCAR and Vehicle Technologies (FCVT) program and the Hydrogen, Fuel Cells, and Infrastructure Technologies (HFCIT) program. In addition, research and development (R&D) on hydrogen production from coal and nuclear energy is carried out in DOE's Office of Fossil Energy (FE) and its Office of Nuclear Energy (NE), and EERE's biomass program pursues work on biomass and biofuels. The Office of Science's Basic Energy Sciences (BES) program is focused on fundamental work in areas such as hydrogen production, hydrogen storage, fuel cell membranes and electrodes, and catalysts. The U.S. Department of Transportation (DOT) also participates in safety-related work.

SCOPE AND FOCUS OF THE COMMITTEE'S REPORT

This being the second biennial report, the committee has focused on assessing progress in each of the research and program management areas as well as the responses of program management to recommendations made in the Phase 1 report. The statement of task directed the committee to

- Review the challenging high-level technical goals and timetables for government and industry R&D efforts in the various technical areas being addressed by the Partnership.
- Review and evaluate progress and program directions since the Phase 1 review toward meeting the Partnership's technical goals, and examine on-going research activities and their relevance to meeting the goals of the Partnership.
- Examine and comment on the overall balance and adequacy of the research and development effort, and on the rate of progress, in light of the technical objectives and schedules for each of the major technology areas.
- Examine and comment, as necessary, on the appropriate role for federal involvement in the various technical areas under development.
- Examine and comment on the Partnership's strategy for accomplishing its goals.
- Review and assess the actions that have been taken in response to recommendations from the Phase 1 review of the program.

Shortly after the committee was formed, Congress asked the NRC to perform a related study that would develop a roadmap of the budgetary resources required to realize a significant percentage of hydrogen-fueled vehicles sold by 2020 in the United States. Accordingly, NRC appointed the Committee on Assessment of Resource Needs for Fuel Cell and Hydrogen Technology to carry out that study. Several members of the committee reviewing Phase 2 of the FreedomCAR and Fuel Partnership—that is, the authors of this report—also serve on the later committee, and one is its chairman. The report from the later study will be published several months after the present report. The two committees have shared information and attempted to achieve consistency to the extent possible and minimize duplication in performing their separate tasks.

This Summary very briefly discusses the technical areas covered in Chapters 3 and 4 and includes the committee's key recommendations. The body of the report contains additional observations, findings, and recommendations on each aspect of the program. The rest of the Summary contains overall comments and briefly addresses crosscutting issues and funding.

OVERALL COMMENTS

The FreedomCAR and Fuel Partnership is well planned, organized, and managed. It is an excellent example of an effective industry/government cooperative effort. There has been significant progress in most areas since the Phase 1 report, and the committee commends management on its thorough and generally receptive responses to the recommendations in that report. The major accomplishments are summarized in Chapter 5, Overall Assessment. Changes, however, take a long time to implement, especially if they require modification of an existing multiyear research contract or a reallocation of funding, which can take up to 2 years. This time factor places a premium on using the best systems analysis tools for program planning and decision making.

There remain many barriers to achieving the objectives of the Partnership (see Chapter 5). These barriers include cost and performance at the vehicle, system, and component levels. To be overcome, some of these barriers will require invention, and others will require new understanding of the underlying science. And, even without these cost and performance barriers, broad economic and social issues might deter so fundamental a transition. Transition to a new, unfamiliar energy carrier, hydrogen, will have to be planned and managed with fact-grounded foresight. The current technology validation program is well conceived and is an important step in addressing those broader issues.

The technical and cost challenges are being addressed by the research sponsored by the Partnership. The committee believes that the expense of this research, if it overcomes these barriers, is justified by its potentially enormous benefits to the nation relative to its use of petroleum. Furthermore, with increased national interest in reducing greenhouse gas emissions, the research efforts of the Partnership are more needed than ever before.

TECHNICAL AREAS

Advanced Combustion Engines and Emission Controls

Internal combustion engines (ICEs) will be the mainstay of the nation's automotive fleet for a very long time, even if the goals of the fuel cell program and the hydrogen infrastructure program are met, enabling fuel cell vehicles to be introduced in large numbers by 2020. Therefore, improvement in the efficiency of these powerplants through combustion research on advanced ICEs is a very important part of the Partnership. This kind of research has provided understanding that allows ICE engines to meet emission constraints and efficiency goals simultaneously. Many experts foresee additional incremental improvements, and there is intense pressure for automobile companies to increase engine efficiency, because it would have an immediate, significant effect on petroleum use. As a result, new findings are quickly translated into large-scale development activities and, if they are successful, will be rapidly deployed by industry. This makes it important that the Partnership's management continually evaluate the ICE research that is being funded to ensure that it is precompetitive.

Recommendation. The Partnership should formulate and implement a clear set of criteria to identify and provide support to ICE combustion and emission control projects that are precompetitive and show potential for improvements well beyond those currently being developed by industry.

Fuel Cells

The development of fuel cells for vehicles and of an infrastructure to deliver hydrogen fuel promises to be one of the most efficient and least polluting ways to power personal transportation vehicles while providing the potential for meeting the Partnership's major goals. Hundreds of fuel cell vehicles are now being built for field tests by auto manufacturers, but these early systems still need significant improvements in durability and cost before mass-produced vehicles can be built and sold. Therefore, an improvement in durability and performance and a reduction in the cost of fuel cell systems remain major goals of the Partnership.

Past R&D has led to important advances, and these continue to occur. Much remains to be done, ranging from basic sciences research to laboratory testing, as well as the validation of results in vehicle tests of complete systems in order to be certain that goals have been met. The breadth and magnitude of these efforts necessitate continual reassessment of the goals and timing of the program elements to assure the appropriate allocation of funds as new knowledge is generated. Many uncertainties remain regarding the likelihood of meeting these goals and timing targets, but the potential benefits of fuel cells and the progress to date certainly justify current spending and increased future spending and budget allocations.

Recommendation. The Partnership should conduct sensitivity analyses on key fuel cell targets to determine the trade-offs and tolerances in engineering specifications allowable while still meeting fuel cell vehicle engineering requirements.

Recommendation. The Partnership should reassess the current allocation of funding within the fuel cell program and reallocate as appropriate, in order to prioritize and emphasize the R&D that addresses the most critical barriers. In particular, the Partnership should give membranes, catalysts, electrodes, and modes of operation the highest priority. It should also

- Place greater emphasis on the science and engineering at the cell level and, from a systems perspective, on integration and subcomponent interactions;
- Reduce research on carbon-based supported catalysts in favor of developing carbon-free electrocatalysts;
- Ensure that Basic Energy Sciences (BES) funding of membranes, catalysts, and electrodes remains a high priority of the program; and
- Apply the go/no-go decision-making process to stationary fuel cell systems initiatives that are not directly related to transportation technologies.

Onboard Hydrogen Storage

Substantially improved techniques for storing hydrogen must be developed to meet the Partnership's goals. Efforts to discover a viable alternative to compressed hydrogen gas are in their very early stages—too early to have confidence in their ultimate success. Therefore, almost all current auto company field test vehicles use 5,000 to 10,000 pound per square inch (psi) (35 to 70 MPa) compressed gas storage.

Meeting the program storage goals almost certainly will require a storage technology as yet undiscovered, making the current search for new storage materials and operating modes appropriate. The funding for this research increased substantially, from $26 million in FY06 to $34.6 million in FY07, although the FY06 appropriation was well below what was requested. The systems analysis techniques being developed should enable the allocation of these funds so that promising approaches are emphasized and progress speeded.

Recommendation. The hydrogen storage program should continue to be supported by the Partnership at a high level since finding a suitable storage material is critical to fulfillment of the vision for the hydrogen economy. Both basic and applied research should be conducted.

Recommendation. The Partnership should rebalance the R&D program for hydrogen storage to shift resources to the more promising approaches as knowledge is gained. The new systems engineering center of excellence (COE) should look at all of the system requirements simultaneously, not just the system weight percent storage goal, and guide this rebalancing.

Recommendation. In the event that no onboard hydrogen systems are found that are projected to meet targets, the Partnership should perform appropriate studies to determine the risks and consequences of relying on pressurized hydrogen storage. It should consider production and delivery issues as well as effects on vehicle performance, safety, and costs.

Electrochemical Energy Storage

Improved battery performance, durability, and cost are critical to gaining more widespread acceptance for hybrid and plug-in hybrid automobiles (including fuel cell hybrids). Very significant progress has been made during the last 2 years, and lithium ion batteries have been developed that can meet several of the FreedomCAR 2010 goals, including weight, volume, and cycle life requirements, with good prospects for meeting the remaining goals as well as the calendar life requirements. New approaches have increased the safety and abuse tolerance of these batteries. Cost is the largest remaining barrier, with estimates of current cost about two times the 2010 goal. Substantial additional research is ongoing to find lower cost materials. The success of this effort will largely determine the viability of these batteries in mass-produced hybrid and plug-in hybrid electric vehicles (PHEVs).

A significant additional breakthrough in battery technology is needed to enable a competitive all-electric automobile that would help meet the FreedomCAR goals. Furthermore, the potential benefits of PHEVs in reducing petroleum consumption have been recognized by the Partnership, yet there seems to be a lack of urgency in finalizing and executing the R&D plan for PHEVs.

Recommendation. The Partnership should conduct a thorough analysis of the cost of the Li ion battery for each application: hybrid electric vehicles (HEVs), PHEVs, battery electric vehicles (EVs), and hydrogen-fueled fuel cell HEVs. The analysis should reexamine the initial assumptions, including those for both market forces and technical issues, and refine them based on recent materials and process costs. It should also determine the effect of increasing production rates for the different systems under development.

Recommendation. The Partnership should significantly intensify its efforts to develop high-energy batteries; in particular it should look for newer higher-specific-energy electrochemical systems within the long-term battery research subactivity and in close coordination with BES.

Recommendation. The Partnership should move forward aggressively with completing and executing its R&D plan for plug-in hybrid electric vehicles.

Electric Propulsion, Electrical Systems, and Power Electronics

HEVs, PHEVs, EVs, and fuel cell vehicles all require electric propulsion and power electronics systems, along with appropriate electronic controllers, to translate electric energy into vehicle propulsion. Improvement in the size, weight, efficiency, and cost of these components is a significant part of the challenge of making such vehicles competitive in the marketplace.

Higher-temperature operation of these components and the integration of power controllers and electric motors to improve the performance of vehicle electric propulsion systems are the most important efforts being supported by the electrical systems and power electronics program, and appropriately so. Higher speed electric motors for vehicle propulsion are being studied in order to reduce their size and weight.

The Partnership supports a wide range of research activities associated with these electrical devices aimed at incrementally improving each of the constituent technologies. Continued improvement in their respective properties is required to help enable viable mass-produced vehicles employing electric propulsion.

Recommendation. The Partnership should conduct a meta-analysis and develop quantitative models to identify fundamental geometric limitations that ultimately set bounds on and lead to the realization of the size, mass, and cost of power converters and electric propulsion systems in relation to the physical properties of materials and processes such as dielectric strength, magnetic saturation, and thermal conductivity. This will allow the various ongoing and future efforts to be benchmarked against the theoretical boundaries of what is possible and enable the establishment of appropriate directions in research goals.

Structural Materials

The Partnership has set a target of a 50 percent reduction in vehicle structural mass with no increase in the cost of the materials involved. From a program management standpoint, either this mass reduction must be achieved or the size and mass of the vehicle powerplant, most other components, and the vehicle's fuel storage capacity will have to be increased.

This Partnership and the earlier Partnership for a New Generation of Vehicles have a long history of research into structural materials for lightweight vehicles that is described in earlier reports. Based upon that work, it is likely that the proposed 50 percent reduction in mass can be achieved. However, it is also quite certain that, within the time frame of the Partnership, this reduction cannot be achieved without incurring a cost penalty. The program management should, accordingly, realistically assess the cost of making the required mass reduction and adjust the cost targets of the other components appropriately. The Phase 1 report recommended reduction of funding, but the lightweight materials programs have continued unabated. In addition, the committee believes that much of the funding currently allocated to the application of lightweight structural materials will not affect this cost penalty significantly and might be better used in other parts of the program.

Recommendation. Based on the 50 percent weight reduction as *a critical goal* and the near-certainty that some (probably significant) cost penalty will be associated with it, the Partnership should develop a materials cost model (even if only an approximation) that can be used in a total systems model to spread this increased cost in an optimal way across other vehicle components.

Recommendation. The materials research funding should largely be redistributed to areas of higher potential payoff, such as high-energy batteries, fuel cells, hydrogen storage, and projects associated with infrastructure issues. However, materials research for projects that show a high potential for enabling near-term, low-cost mass reduction should continue to be funded.

Hydrogen Production, Delivery, and Dispensing

Hydrogen Fuel Pathways

The Partnership envisions a gradual transition from petroleum-based fuel to hydrogen as the main energy carrier for transportation vehicles. There are many pathways that such a transition might follow, and each needs to be analyzed and understood.

The transition envisioned is likely to take place in complex ways over substantially more than a decade as the population of fuel cell vehicles grows. It is reasonable to expect hydrogen initially to come from existing centralized production facilities and to be distributed by tube trailer or liquid carrier. These supplies are likely to be supplemented, increasingly, by distributed generation in service station forecourts, using steam reforming of widely distributed natural gas, or by water electrolysis powered by the electric grid, perhaps during off-peak periods. Such methods are likely to continue to be used until fuel demand in populated areas is sufficient to justify the distribution by pipeline of hydrogen from centralized sources and produced in several different ways. Even then, some remote areas are likely to continue to be supplied by the methods used early in the transition.

With all of these potential pathways, more extensive scenario analysis of the transition and emergence of mature hydrogen-fueled systems is needed to enable us to understand the most critical factors in production and delivery as the market develops.

Recommendation. DOE should continue its studies of the transition to hydrogen, extending them to 2030-2035, when the number of hydrogen vehicles in use could increase rapidly, and use the results of these studies as a basis for evaluating the potential roles of different transitional supplies of hydrogen fuel as demand increases substantially, including both forecourt production at the fueling station and centralized production using the most cost-effective means of distributing the hydrogen.

Hydrogen Production and Delivery

Hydrogen is currently produced in large quantities in centralized plants and is widely distributed in both gaseous and liquid forms for a variety of uses. However, the challenges of producing it and delivering it in appropriate quantities over several years to a growing transportation vehicle fleet are significant. Given concerns about carbon dioxide emissions and the need for ubiquitous delivery points and safety, many scenarios and a variety of raw materials and production processes must be considered and analyzed.

The Partnership's production goals for vehicular hydrogen assume that it will come from a diverse set of feedstocks. Natural gas reforming is the most straightforward method of distributed hydrogen production at local service stations during the transition period. However, this process will result in greenhouse gas emissions and increased imports of natural gas, and because its space requirements (and that of other distributed generation schemes) could limit its use, it will need further study.

The development of carbon capture and sequestration (CCS) technology in FE will pace the possible economic production of hydrogen from coal. DOE has made important progress in identifying the potential supplies of biomass for conversion to hydrogen and other fuels. However, extensive research, development, and demonstration work on biomass production and conversion to hydrogen must be completed and water and land issues addressed to determine the amount of hydrogen that can be sustainably produced at a competitive cost. Basic research is required to determine the feasibility of new processes for producing hydrogen with nuclear reactor heat, and more research is needed to enhance electrolysis technology for splitting water.

Unlike the distribution of gasoline or diesel fuel, hydrogen distribution will consume substantial energy and incur significant costs. These energy losses and costs need to be considered in choosing appropriate delivery methods during various stages of the transition.

Recommendation. DOE should put more emphasis on the space requirements for forecourt hydrogen generation by studying ways to minimize these requirements.

Recommendation. DOE should conduct a systematic review of the CCS program as it affects the schedule for and program assumptions about hydrogen production from coal. This review should identify indicators of incipient program slippage and, through systems analysis, the program consequences of possible delays, leading to recommendations for management actions that would compensate for these delays.

Recommendation. The Partnership should increase funding for electrolysis efforts to advance the technology, demonstrations, and systems integration. BES should support, as appropriate, fundamental research in catalysts, membranes, and coatings as well as in new concepts.

Recommendation. DOE should undertake a systems study to assess the relative importance of barriers to biomass production, availability, transportation, and conversion to hydrogen in order to identify the areas that most affect commercial availability, giving them priority attention in the program. This study should address technical barriers already identified, including their impact on the environment, and help define policies for land and water use and government-sponsored commercial incentives that would stimulate commercial expansion of the biomass options.

Recommendation. DOE should involve the energy partners in all biomass programs related to conversion to hydrogen or hydrogen carriers as early in the programs as possible.

Recommendation. DOE should increase funding for the delivery and dispensing program to meet the market transition and sustained market penetration time frames. If DOE concludes that a funding increase is not feasible, the program should be focused on the pipeline, liquefaction, and compression programs, where a successful, if only incremental, short-term impact could be significant for the market transition period.

CROSSCUTTING ISSUES

Safety

Safety is recognized as a critical, overarching factor throughout all of the Partnership activities. One part of the effort is assuring that all activities involving hydrogen are carried out in a safe manner and that lessons learned are put into standard practice. Another part is research on the safe use of hydrogen for fueling and operating vehicles and developing appropriate codes and standards.

The safety activity was funded at well below requested levels until 2007, when its budget increased to $13.8 million. This money is helping to support many organizations developing vehicle and component standards, work on fueling station design, fast fueling to 70 MPa (10,000 psi), and the development of hydrogen quality standards. There is also an extensive program on unintentional releases of hydrogen, its behavior, safety sensors, and materials compatibility with hydrogen.

The task of developing adequate safety codes and standards is immense and unlikely to be completed by 2010. Getting new codes and standards adopted can easily take 7-10 years, and lack of appropriate regulations could impede the introduction of hydrogen vehicles into the marketplace. This underscores the importance of developing the underlying knowledge and providing it to codes and standards organizations as soon as possible.

Recommendation. DOE should establish a program to address all end-to-end safety aspects as well as codes and standards. Such a program could be viewed as an extension of the current quantitative risk analysis activity, which is focused on filling stations. This task should be adequately funded and expanded. The priority for expansion should go to (1) the vehicle and (2) the fuel distribution system.

Recommendation. The Department of Transportation (DOT), including all relevant entities within it, should develop a long-range, comprehensive hydrogen safety plan with budget estimates and milestones to 2015. The milestones developed in this plan should be integrated into the milestones and roadmap of the codes and standards technical team.

Technical Validation (Learning Demonstration Program)

This program collects data from fuel-cell-powered vehicles being driven on public highways and from hydrogen refueling stations located in a variety of environments around the country. The data collection process is well conceived and is establishing a credible metric for the state of the art for hydrogen supply and fuel cell vehicle systems. Results from this program are pooled and shared and are being used effectively to guide the technical teams and analysis efforts and to set priorities for the overall program. The second generation of vehicles now being put into the program by auto manufacturers will validate overall progress under real-world conditions. This program is an essential way for the Partnership to learn about the real-world performance of the technologies it is developing.

Recommendation. DOE should continue to disseminate the results of the technical validation activity to supporting organizations outside the Partnership in order to promote widespread innovation and competition. DOE management needs to systematically evaluate the information being generated by each project to determine when the project should be terminated. On the other hand, DOE management should not prematurely drop support for the overall technical validation and learning demonstrations, because their importance cannot be overemphasized. DOE and the Partnership should develop a long-range plan for technology validation that continues until at least 2015.

Recommendation. DOE management should maintain adequate support for technical validation as it is essential to the overall Partnership. This support should be balanced and cover both the vehicles themselves and the fuel infrastructure needed. To achieve the rapid learning that the overall project requires, DOE should also keep the validation activities focused on their primary purpose—the accumulation, analysis, and dissemination of experience from the field. Safety should be stressed throughout the learning demonstration program, because an accident early on could attract publicity out of proportion to its true consequences.

Decision Making and Strategic Planning

Management of the FreedomCAR and Fuel Partnership is a complex task because of the Partnership's breadth, its technological sophistication, and its need for ongoing commitments from automobile companies and energy companies, as well as the federal government. The organization of the Partnership provides for the involvement of appropriate people to perform the needed tasks, all of the way from research scientists to those providing strategic direction. The earlier Partnership for a New Generation of Vehicles (PNGV) proved that this basic structure is effective.

A program of this scope requires effective decision-making and strategic planning processes to ensure its objectives are appropriately focused and that good progress is being made. Program management requires a variety of system analysis tools applicable to accomplish this, and during the 14 years of the two programs, several such tools have been developed (Chapter 2). Individual activities of the program have used these tools effectively to explore various scenarios and to set and modify goals. What is still needed, however, is a tool to quantitatively assess how the various technology options being pursued will impact the overall goals of reducing petroleum consumption, air pollutants, and greenhouse gas emissions when the options are deployed in the marketplace. Such a tool is under development and is scheduled for completion in 2008. It should be used in conjunction with panels of outside experts who can give the program managers and the Executive Steering Group advice on the validity of the models, technical risk and market risk, the role of possible market interventions, and the best role for DOE vis-à-vis the private sector in the development of various technologies. The importance of reducing U.S. petroleum consumption and greenhouse gas emissions is ever more recognized, which makes a high-level review and assessment of this program and its effectiveness very timely.

Recommendation. DOE should accelerate the development and validation of tools that can be used to model propulsion system and vehicle technologies and fuels and determine their potential impact on the overall Partnership goals of reducing petroleum use and air pollutant and greenhouse gas emissions. Sensitivity analyses, from worst case to best case scenarios, should be performed to assess these impacts. Models, input data, and assumptions should be independently reviewed to validate and refine the models and lend credibility to the conclusions derived from them.

Recommendation. The Partnership should use its technical and cost systems analysis capabilities to provide an essential component to manage its program, to assess progress in meeting technical and cost targets, to examine the impact of not meeting these targets, to adjust program priorities, and to make go/no-go decisions.

In the committee's judgment, the activities being pursued by the Partnership have a critical role to play in carving out a sustainable path for the U.S. transportation system. As anticipated at the inception of the program, this path will include a wide variety of new and improved propulsion systems, vehicle technologies, and fuels. The program goals are very challenging, and the importance of achieving them becomes more pressing each year. The committee believes that it is time to step back and, with the knowledge already gained, engage in a strategic review in the context of other ongoing domestic and international activities focused on vehicle and fuel technologies. The leadership of the Partnership is eminently qualified to oversee this review with the goal of ensuring that the activities undertaken in the next few years are adequate to meet the challenges now evident.

Recommendation. The Executive Steering Group of the FreedomCAR and Fuel Partnership should establish a high-level planning group to develop a strategic plan appropriate for the next phase of the nation's collaborative R&D program for vehicle and fuels technology.

Program Balance and Funding

The total FY07 budget for the hydrogen-related activities that make up the Hydrogen Fuel Initiative (hydrogen technologies and fuel cells) within DOE is about $274 million, and the total funding of relevance to the charter of the committee is about $401 million (see Chapter 5). Overall program funding is consistent with recommendations of previous in-depth studies and is consistent with the President's commitment of $1.7 billion over 5 years (FY04 to FY08).

The committee has proposed an overall assessment of the program goals in each technology area and expects this to provide a better basis for judging the adequacy of funding in each area as the program moves forward beyond FY08. The committee also has noted some specific areas that should be expanded (i.e., electrolysis; hydrogen delivery/distribution; forecourt designs with minimum space requirements; fuel cells; safety, codes and standards) and has indicated that the proposed reduction in FY08 funds for the technology validation program be restored. Structural materials research is the only endeavor where the committee feels some funds should be reallocated to more critical projects.

The committee also recommended that a strategic planning assessment be performed to ensure that the program activities are adequate to achieve its goals, which are of great strategic importance to the United States. Such an assessment should be part of the recommendation to develop a broad forward-looking strategic plan and would provide a basis for determining longer-term funding needs, given the importance of U.S. energy security and reductions in greenhouse gas emissions. In addition, the ongoing NRC study on hydrogen resources will certainly provide useful information on priorities of the overall effort to develop a transportation system that includes hydrogen-powered vehicles.

The committee notes that the congressional practice of earmarking funds has severely restricted the ability of DOE to effectively manage some parts of its program. There is concern about insufficient funding in three other areas as well: the DOE CCS program, the DOT hydrogen safety program, and the hydrogen from biomass activity, but these concerns were not investigated in detail by the committee.

1

Introduction

BACKGROUND

The increased rate at which the world's economies are becoming globalized has brought with it an increased demand for energy. Projections of growth in energy use for the next 30 years suggests that the United States, as well as the rest of the world, will be challenged to supply the energy demanded by these economies (NPC, 2007). All sectors of the economy will be affected. Mobility systems account for approximately 28 percent of the total U.S. energy use and approximately 67 percent of U.S. petroleum consumption (EIA, 2005). Consequently, diversifying the energy carriers used in mobility systems and developing new sources for them will be an important component of the U.S. energy situation and are important national issues. Furthermore, concerns about climate change and reducing greenhouse gas emissions have been receiving extensive attention from the Congress, the states, the Supreme Court (on the role of the Environmental Protection Agency in regulating greenhouse gas emissions), and the Intergovernmental Panel on Climate Change. The use of hydrogen in the transportation sector has the potential to reduce greenhouse gas emissions from that sector.

As President Bush said in his 2003 State of the Union address, hydrogen, as an energy carrier, would have many advantages if it could be developed for the mobility market. However, the challenges of doing so are great. The FreedomCAR and Fuel Partnership was established to address these challenges and advance the technology enough so that a decision on the commercial viability of hydrogen vehicles can be made by 2015. This report reviews the status and progress of this Partnership.

The U.S. Department of Energy (DOE) has been involved for about 30 years in research and development (R&D) programs related to advanced vehicular technologies and alternative transportation fuels. During the 1990s, much of this R&D was conducted under the Partnership for a New Generation of Vehicles (PNGV) program, which was formed between the federal government and the auto industry's U.S. Council for Automotive Research (USCAR).[1] Building on the PNGV program, in January 2002, the Secretary of Energy and executives of DaimlerChrysler, Ford, and General Motors announced a new government-industry partnership between DOE and USCAR called FreedomCAR, with CAR standing for Cooperative Automotive Research. In September 2003, FreedomCAR was expanded to also include five large energy companies—BP America, Chevron Corporation, ConocoPhillips, ExxonMobil Corporation, and Shell Hydrogen (U.S.)—to address issues related to supporting the fuel infrastructure. The expanded partnership is called the FreedomCAR and Fuel Partnership.[2] The long-term goal of the Partnership is to "enable the full spectrum of light-duty passenger vehicle classes to operate completely free of petroleum and free of harmful emissions while sustaining the driving public's freedom of mobility and freedom of vehicle choice" (DOE, 2004a; DOE, 2004b, p. 1-6).

The Partnership addresses the development of advanced technologies for all light-duty passenger vehicles: cars, sport utility vehicles (SUVs), pickups, and minivans. It also addresses technologies for hydrogen production, distribution, dispensing, and storage. The Partnership started with a presidential commitment to request $1.7 billion over 5 years (FY04 to FY08), with appropriations thus far of about $243 million, $307 million, and $339 million for FY04, FY05, and FY06, respectively. The FY07 Continuing Resolution resulted in funding of about $401 million. The FY08 presidential budget request is for about $436 million (see Chapter 5). Funding for research, development, and demonstration activities goes to universities, the national laboratories, and private companies. Especially in the

---

[1]USCAR, which predated PNGV, was established by Chrysler Corporation, Ford Motor Company, and General Motors Corporation. Its purpose was to support intercompany, precompetitive cooperation so as to reduce the cost of redundant R&D, especially in areas mandated by government regulation, and to make the U.S. industry more competitive with foreign companies. Chrysler Corporation merged with Daimler Benz in 1998 to form DaimlerChrysler. In 2007, DaimlerChrysler divested itself of a major interest in the Chrysler Group, and Chrysler LLC was formed; DaimlerChrysler will be renamed Daimler AG.

The PNGV sought to significantly improve the nation's competitiveness in the manufacture of future generations of vehicles, to implement commercially viable innovations emanating from ongoing research on conventional vehicles, and to develop vehicles that achieve up to three times the fuel efficiency of comparable 1994 family sedans (NRC, 2001; PNGV, 1995; The White House, 1993).

[2]In February 2003, before the announcement of the FreedomCAR and Fuel Partnership, the President announced the FreedomCAR and Hydrogen Fuel Initiative to develop technologies for (1) fuel-efficient motor vehicles and light trucks, (2) cleaner fuels, (3) improved energy efficiency, and (4) hydrogen production and a nationwide distribution infrastructure for vehicle and stationary power plants, to fuel both hydrogen internal combustion engines (ICEs) and fuel cells (DOE, 2004a). The expansion of the FreedomCAR and Fuel Partnership to include the energy sector after the announcement of the initiative also supports the goal of the Hydrogen Fuel Initiative.

case of development activities, projects are often cost shared between the private sector and the federal government (see Chapter 5 for further discussion).

The Partnership plays an important role in the planning, pursuit, and assessment of high-risk, precompetitive R&D for many of the needed vehicle and fuel technologies. Federal funds enable this work to move forward. The Partnership also serves as a communication mechanism for the interested players, including government, the private sector, the national laboratories, universities, the public, and others.

In late 2006 the National Research Council (NRC) formed the Committee on Review of the FreedomCAR and Fuel Research Program, Phase 2 (see Appendix B for biographical information on the members.) Its report represents the second review by the NRC of the research program of the Partnership. The first review was conducted during 2004 and 2005 and resulted in a report issued in the fall of 2005 (NRC, 2005). (The first review will be referred to as the Phase 1 review and/or report.)

GOALS AND TARGETS

The long-term goal of the Partnership is to enable the transition to light-duty passenger vehicles that operate free of petroleum and free of harmful emissions (DOE, 2004b). Starting to reduce the nation's dependence on imported petroleum is central to this goal. The current plan envisions a pathway starting with more fuel-efficient ICEs and hybrid electric vehicles (HEVs), including plug-in HEVs (PHEVs), potential use of all-electric-drive vehicles, and, ultimately, addition of an infrastructure for supplying hydrogen fuel for fuel-cell-powered vehicles (DOE, 2004b).

To address the technical challenges associated with this envisioned pathway, the Partnership has established quantitative technology and cost goals for 2010 and 2015 in eight areas:

- ICEs (both petroleum- and hydrogen-fueled),
- Fuel cell power systems,
- Fuel cells,
- Hydrogen storage systems,
- Energy storage systems for hybrid vehicles,
- Hydrogen production and delivery systems,
- Electric propulsion systems, and
- Materials for lightweight vehicles.

These goals and the research related to their attainment will be discussed later in this report. Technical teams, as noted in the next section, "Organization of the Partnership," specify and manage technical and crosscutting needs of the Partnership.

ORGANIZATION OF THE PARTNERSHIP

The Partnership consists of a number of oversight groups and technical teams that have participants from government and industry (see Figure 1-1). The Executive Steering Group, which is responsible for the governance of the Partnership, is made up of the DOE assistant secretary for energy efficiency and renewable energy (EERE) and a vice-presidential-level executive from each of the Partnership companies. It meets as needed. The FreedomCAR Operations Group, made up of DOE program managers and directors from USCAR member companies, is responsible for directing the technical teams and prioritizing research issues. The Fuel Operations Group, made up of DOE program managers and energy company directors, is responsible for the direction of the fuel technical teams. Periodically, the FreedomCAR Operations Group and the Fuel Operations Group hold joint meetings to coordinate fuel and power plant issues and identify strategic or policy issues that warrant attention by the Executive Steering Group (DOE, 2004c).

The Partnership has formed 11 industry-government technical teams responsible for R&D on the candidate subsystems (see Figure 1-1). Most of these technical teams focus on specific technical areas, but some, such as codes and standards and vehicle systems analysis, focus on crosscutting issues. A technical team consists of scientists and engineers with technology-specific expertise from the USCAR member companies, energy partner companies, and national laboratories,

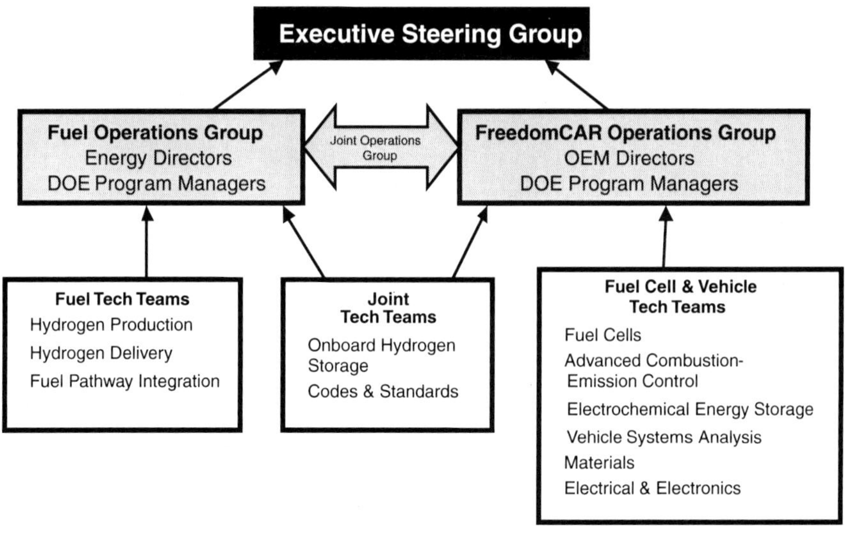

FIGURE 1-1 FreedomCAR and Fuel Partnership organizational structure. SOURCE: E.J. Wall and J. Milliken, "Overview of the FreedomCAR and Fuel Partnership," Presentation to the committee on March 1, 2007.

as well as DOE technology development managers. They may come from other federal agencies if approved by the appropriate operations group(s). A technical team is responsible for developing R&D plans and roadmaps, reviewing research results, and evaluating technical progress toward meeting established research goals (DOE, 2004c). Its discussions are restricted to nonproprietary topics.

Fuel cell and vehicle technical team members come from the USCAR partners and DOE. They handle fuel cells, advanced combustion and emissions control, systems engineering and analysis, electrochemical energy storage, materials, and electrical systems and power electronics. The three fuel technical teams address hydrogen production, hydrogen delivery, and fuel/vehicle pathway integration, each of which has members from the energy companies and DOE. There are two joint technical teams connecting the fuel teams and the vehicle teams: an onboard hydrogen storage team and a codes and standards team.

At DOE, primary responsibility for the FreedomCAR and Fuel Partnership rests with EERE.[3] The two main program offices within EERE that manage the Partnership are the FreedomCAR and Vehicle Technologies (FCVT) program and the Hydrogen, Fuel Cells, and Infrastructure Technologies (HFCIT) program.

The FCVT program has the following specific goal: to support "R&D that will lead to new technologies that reduce our nation's dependence on imported oil, further decrease vehicle emissions, and serve as a bridge from today's conventional power trains and fuels to tomorrow's hydrogen-powered hybrid fuel cell vehicles" (DOE, 2004b, p. ES-2). The FCVT also includes the 21st Century Truck Partnership.[4]

The FreedomCAR and Fuel Partnership activities in the FCVT program are organized into these areas:

- Vehicle systems analysis and testing to provide an overarching vehicle systems perspective to the technology R&D subprograms and other activities in the FCVT and HFCIT programs;
- Advanced energy-efficient, clean ICE power trains using various petroleum and nonpetroleum-based fuels, including hydrogen;
- Electric energy storage technologies (batteries and ultracapacitors);
- Advanced power electronics and electric machines;

[3]EERE has a wide variety of technology R&D programs and activities related to renewable energy technologies, such as the production of electricity from solar energy or wind and the production of fuels from biomass, to the development of technology to enhance energy efficiency, whether for vehicles, appliances, buildings, or industrial processes. It also has programs on distributed energy systems (see Appendix A for an EERE organization chart).

[4]DOE supports several other programs related to the goal of reducing dependence on imported oil. The 21st Century Truck Partnership supports R&D on more efficient and lower emission commercial road vehicles. The NRC Committee on Review of the 21st Century Truck Partnership Program is reviewing that program.

- Materials technology for lightweight vehicle structures and for propulsion system components, including power electronics and ICEs; and
- Fuels technologies that enable current and emerging advanced ICEs and emission control systems to be as efficient as possible while meeting future emission standards and that reduce reliance on petroleum-based fuels.

The HFCIT program directs activities in hydrogen production, storage, and delivery and integrates them with transportation and fuel cell development activities. The proton exchange membrane (PEM) fuel cell R&D is undertaken in the HFCIT program, which is focused on

- Overcoming technical barriers through R&D on hydrogen production, delivery, and storage technologies, as well as fuel cell technologies for transportation, distributed stationary power, and portable power applications;
- Addressing safety concerns and developing model codes and standards;
- Validating and demonstrating hydrogen fuel cells in real-world conditions; and
- Educating key stakeholders whose acceptance of these technologies is critical to their success in the marketplace (DOE, 2004a,b).

The manager of HFCIT is the overall DOE hydrogen technology program manager.

Some activities related to the HFCIT program focus are not within EERE. The Office of Fossil Energy (FE) supports the development of technologies to produce hydrogen from coal and to capture and sequester carbon. The Office of Nuclear Energy (NE) supports research into the potential use of high-temperature nuclear reactors to produce hydrogen, while the Office of Science (SC) supports fundamental work on new materials to store hydrogen; catalysts; fundamental biological or molecular processes for hydrogen production; fuel cell membranes; and other related basic science areas (DOE, 2004d,e). Within EERE there is also an Office of Biomass Energy, which is not part of the FreedomCAR and Fuel Partnership. However, biomass is of interest to the Partnership both as one possible source of hydrogen and as part of a strategy to diversify energy sources for the transportation sector, so there is cooperation between the Partnership and the biomass program.

RECENT INITIATIVES

Since the Phase 1 review by the NRC and the ensuing 2005 report, a number of external developments have occurred that may affect the program (NRC,

2005). There has been an increasing interest on the part of both the Congress and the administration in the security implications of U.S. dependence on imported energy, especially petroleum, as well as the issues of global warming and emissions of greenhouse gases. As a result, President Bush has called for reducing gasoline consumption by 20 percent in 10 years (by 2017) through a combination of alternative fuels and reform of the Corporate Average Fuel Economy (CAFE) standards for cars. He has called for the production of 35 billion gallons of fuel from renewable sources and other alternative fuels as part of this effort to reduce gasoline consumption. Congress has supported expanded production of fuel ethanol, which increased rapidly during the past few years and reached about 5.4 billion gal/yr in 2006, and is providing incentives for much more expansion.[5] Although ethanol production in the United States is now mostly from corn, eventually, ethanol is expected to be produced from cellulose (e.g., grasses, woody plants, and agricultural and wood wastes). Such processes still require substantial R&D to be successful. Other potential alternative fuels include gasoline or diesel liquids derived from coal or oil shale. Many alternatives are being explored, but which fuels and to what extent they will be able to enter the marketplace by 2017 remains very uncertain.

In addition, there are numerous bills in Congress aimed at achieving significant reductions in greenhouse gas emissions. If passed, these bills may create incentives to either improve the fuel economy of vehicles or stimulate the adoption of fuels that produce less greenhouse gases than do gasoline and diesel fuel.

There has also been increasing interest in PHEVs, which would contain an energy conversion device, such as an ICE, and a battery that could be charged from the electric grid when not in use. Depending on the battery capacity and control logic, a version of this car could be driven between 20 to 40 miles on battery power alone, which is the distance many people drive to work every day. A cost-effective, durable battery of adequate capacity would enable the electric grid to supply a significant part of the energy for U.S. vehicles. Since virtually no petroleum is used to produce electricity in the United States, this would reduce demand for petroleum in the transportation sector. However, depending on the mix of fuels used to supply electricity for such vehicles, this could lead to increased natural gas imports and consumption of coal, with implications for greenhouse gas emissions. The Energy Policy Act of 2005 called for a research program on such vehicles as well as flexible-fuel vehicles (e.g., vehicles that can use gasoline or ethanol or a mixture of both). The President's Advanced Energy Initiative[6] called for the development of advanced battery technologies that would enable a plug-in hybrid vehicle to go 40 miles on battery power alone. The Phase

[5]See the Renewable Fuels Association Web site at <http:www.ethanolrfa.org/industry/statistics/#D>.

[6]The Advanced Energy Initiative report can be found at <http://www.whitehouse.gov/stateoftheunion/2006/energy/energy_booklet.pdf>.

1 review of the Partnership also called for increased research on such high-energy storage batteries.

This increased interest on the part of the public, Congress, and the administration in reducing petroleum use, and hence energy imports and greenhouse gas emissions, could further stimulate interest in the development of hydrogen-fueled vehicles. But it could also shift funding to biofuels, alternative liquid fuels, and plug-in hybrids, creating competition for hydrogen-fueled fuel cell vehicles.

VEHICLE AND FUEL TECHNOLOGIES

The Phase 1 review of the Partnership contains some general discussion of the importance of linking vehicles, fuels, and infrastructure to ensure that the impacts on the commercial market will be significant and widespread. (That discussion will not be repeated here, and the reader is referred to the Phase 1 report for that background [NRC, 2005, Chapter 1].) Successful examples of new fuels include the introduction of unleaded gasoline in 1971 and the introduction of reformulated gasoline in the 1990s. But efforts to introduce alternative fuels such as methanol, ethanol, and compressed natural gas on a wide scale have all foundered, in part owing to economics. Alcohol fuels, such as 85 percent methanol (M85) or 85 percent ethanol (E85), work well in vehicles designed to accept them, and although there are several million vehicles on the road that can use these fuels, no extensive fueling infrastructure has followed suit. Compressed natural gas (CNG) vehicles have also enjoyed limited success. They are mainly found in fleets and in niche markets. This need for both the acceptance of new vehicle technology that relies on nontraditional fuel and the widespread availability of that fuel in the marketplace is why the Partnership supports R&D for both vehicles and fuels. The Partnership seeks ultimately to enable the widespread deployment of fuel cell vehicles fueled by convenient, competitively priced hydrogen, and it is structured to address the obvious barriers to achieving this goal for both the fuel cell vehicle and the fuel production and delivery systems.

Hydrogen represents a totally new fuel for the transportation sector, and a totally new infrastructure will have to be put in place. This creates a chicken-and-egg situation. Even if successful and cost-competitive fuel cell vehicles are developed, they could not be sold in great numbers if there were no fuel infrastructure. Likewise, an extensive hydrogen fuel infrastructure cannot be economically justified to service the first few fuel-cell-powered vehicles that might be built. *The Hydrogen Economy* emphasized the importance of distributed production of hydrogen, e.g., using natural gas and the existing infrastructure to produce hydrogen at fueling stations, or using renewable energy—for example, wind to electric systems—to generate hydrogen through electrolysis at the fueling stations (NRC/NAE, 2004). Generating hydrogen at the fueling station would avoid the need to initially install a vast hydrogen distribution infrastructure. DOE has focused significant efforts on this transition concept, as discussed in Chapter 4.

Even in the most optimistic scenario postulated in *The Hydrogen Economy*, only 10 percent of new vehicles and 6 percent of the total miles traveled in 2024 are accounted for by hydrogen-fueled fuel cell vehicles (NRC/NAE, 2004). The remaining 90 percent of new vehicles are projected to be conventionally powered vehicles, either hybrid or nonhybrid. Consequently, by far the greatest contribution to reduced energy use and emissions by and from the U.S. vehicle fleet over the next 20 years and beyond will come from continued improvement in ICEs, HEVs, and their fuels.

To reduce transportation fuel use, current industry-wide efforts to improve the efficiency of ICEs and to further develop the corresponding fuels must continue or, better, accelerate. This is true regardless of the degree to which HEV power trains proliferate or whether advanced diesel engines achieve customer acceptance and meet emission standards. The urgency of this task is amplified by the reality that even with approximately 16 million new vehicles sold in the United States every year, it takes almost 15 years to turn over the national fleet of roughly 225 million vehicles.

While much of the Partnership activity is devoted to fuel cell vehicles and hydrogen fuel, further improvement in conventional ICEs and HEVs could contribute significantly to the goals of energy independence and reduced carbon emissions and should benefit from continued collaboration between industry engineers and the DOE national laboratories in this area. The status of Partnership efforts to develop ICEs and emission control technologies is discussed in Chapter 3.

COMMITTEE APPROACH AND ORGANIZATION OF THIS REPORT

The statement of task for this committee is as follows:

- Review the challenging high-level technical goals and timetables for government and industry R&D efforts, which address such areas as (1) integrated systems analysis; (2) fuel cell power systems; (3) hydrogen storage systems; (4) hydrogen production and distribution technologies necessary for the viability for hydrogen-fueled vehicles; (5) the technical basis for codes and standards; (6) electric propulsion systems; (7) electric energy storage technologies; (8) lightweight materials; and (9) advanced combustion and emission control systems for internal combustion engines (ICEs).
- Review and evaluate progress and program directions since the Phase 1 review toward meeting the Partnership's technical goals, and examine ongoing research activities and their relevance to meeting the goals of the Partnership.
- Examine and comment on the overall balance and adequacy of the research and development effort, and the rate of progress, in light of the technical objectives and schedules for each of the major technology areas.
- Examine and comment, as necessary, on the appropriate role for federal involvement in the various technical areas under development.
- Examine and comment on the Partnership's strategy for accomplishing its goals, which might include such issues as (1) program management and organization; (2) the process for setting milestones, research directions, and making go/no-go decisions; (3) collaborative activities needed to meet the program's goals (e.g., among the various offices and programs in DOE, DOT, USCAR, universities, the private sector, and oth-

ers); and (4) other topics that the committee finds important to comment on related to the success of the program to meet its technical goals.

- Review and assess the actions that have been taken in response to recommendations from the Phase 1 review of the program. Write a report documenting its conclusions and recommendations.

The committee met three times to hear presentations from DOE and industry people involved in the management of the Partnership and to discuss insights gained from both the presentations and written material gathered by the committee and a fourth time to review drafts of the report sections (see Appendix C for a list of committee meetings). The committee established subgroups to investigate specific technical areas and formulate questions for the program leaders to answer. The subgroups also met with the Partnership technical team leaders to clarify answers to questions and better understand the team dynamics, and several committee members visited the General Motors Honeye facility in New York to view its fuel cell vehicle developments.

Concurrently with this review, the NRC is engaged in another related study being undertaken by the Committee on Resource Needs for Fuel Cell and Hydrogen Technologies. That committee is charged with creating ". . . a budget roadmap of the resources required to realize a significant percentage of hydrogen-fueled vehicles sold by 2020" and will publish its report soon after this report has been published. Coordination between the efforts has been achieved by having two individuals become members of both committees, having members of both committees attend meetings of the other committee, and having informal telephone conversations between the technical experts on this committee and group leaders on the other committee.

The Summary presents the committee's main conclusions and recommendations. This chapter (Chapter 1) provides background on the FreedomCAR and Fuel Partnership, on its organization, and on the dual nature—vehicle development and fuel development—of the program. Chapter 2 examines the important crosscutting issues that the program is facing. Chapter 3 looks more closely at R&D for vehicle technology, and Chapter 4 examines R&D for hydrogen production, distribution, and dispensing. Finally, Chapter 5 presents an overall assessment.

REFERENCES

Department of Energy (DOE). 2004a. *Hydrogen, Fuel Cells and Infrastructure: Multi-Year Research, Development and Demonstration Plan*. DOE/GO-102003-1741. Washington, D.C.: U.S. Department of Energy, Office of Energy Efficiency and Renewable Energy. Available on the Web at <http://www.eere.energy.gov/hydrogenandfuelcells/mypp/>.

DOE. 2004b. *FreedomCAR and Vehicle Technologies Multi-Year Program Plan*. Washington, D.C.: U.S. Department of Energy, Office of Energy Efficiency and Renewable Energy. Available on the Web at <http://www.eere.energy.gov/vehiclesandfuels/resources/fcvt_mypp.shtml>.

DOE. 2004c. *Partnership Plan. FreedomCAR & Fuel Partnership*. Washington, D.C.: U.S. Department of Energy, Office of Energy Efficiency and Renewable Energy. Available on the Web at <http://www.eere.energy.gov/vehiclesandfuels/pdfs/program/fc_fuel_partnership_plan.pdf>.

DOE. 2004d. *Basic Research Needs for the Hydrogen Economy: Report of the Basic Energy Sciences Workshop on Hydrogen Production, Storage, and Use,* May 13-15, 2003. Washington, D.C.: U.S. Department of Energy, Office of Science. Available on the Web at <http://www.er.doe.gov/production/bes/hydrogen.pdf>.

DOE. 2004e. *Hydrogen Posture Plan: An Integrated Research, Development and Demonstration Plan*. Washington, D.C.: U.S. Department of Energy. Available on the Web at <http://www.eere.energy.gov/hydrogenandfuelcells/pdfs/hydrogen_posture_plan.pdf>.

Energy Information Administration (EIA). 2005. *Annual Energy Review 2005,* Washington, D.C. Available on the Web at <http://tonto.eia.doe.gov/FTPROOT/multifuel/038405.pdf>.

National Petroleum Council (NPC). 2007. Facing the hard truths about energy: A comprehensive view to 2030 of global oil and natural gas. Executive summary, July 18, 2007. Available on the Web at <http://downloads.connectlive.com/events/npc071807/pdf-downloads/Facing_Hard_Truths-Executive_Summary.pdf>.

National Research Council (NRC). 2001. *Review of the Research Program of the Partnership for a New Generation of Vehicles, Seventh Report*. Washington, D.C.: National Academy Press.

NRC. 2005. *Review of the Research Program of the FreedomCAR and Fuel Partnership, First Report*. Washington, D.C.: The National Academies Press.

National Research Council/National Academy of Engineering (NRC/NAE). 2004. *The Hydrogen Economy: Opportunities, Costs, Barriers, and R&D Needs*. Washington, D.C.: The National Academies Press.

PNGV (Partnership for a New Generation of Vehicles). 1995. *Partnership for a New Generation of Vehicles Program Plan* (draft). Washington, D.C.: U.S. Department of Commerce, PNGV Secretariat.

The White House. 1993. *Historic Partnership Forged with Automakers Aims for Threefold Increase in Fuel Efficiency in as Soon as Ten Years*. Washington, D.C.: The White House.

2

Major Crosscutting Issues

This chapter addresses the main crosscutting issues that the committee identified in its review of the FreedomCAR and Fuel Partnership. Some of these issues deal with the broader context affecting the successful adoption into the marketplace of the technologies under development. The committee recommendations are intended to help the Partnership to progress more rapidly and increase its chances of success. Specific technical areas are discussed in Chapters 3 and 4.

STRATEGIC PLANNING AND DECISION MAKING

Background

The FreedomCAR and Fuel Partnership is an R&D program that focuses on critical transportation technology and fuels challenges, which, if successfully met, could significantly lower U.S. petroleum consumption and greenhouse gas emissions. It is a major program, funded and managed by the federal government (DOE), three U.S. auto companies, and five petroleum companies. The individual technical teams work primarily at the vehicle component level and on the production, distribution, and delivery of hydrogen. To these are added a vehicle systems analysis technical team and a fuel pathway integration technical team (see Figure 1-1). This organizational structure recognizes the need for project activities that focus on individual technical issues, as well as on the characteristics of the integrated vehicle system and the total fuel system. In addition, there is a broader strategic perspective, which the Executive Steering Group provides. The system integration and performance issues require a systems approach at several levels, necessitating a variety of systems analysis tools.

Several such tools are now available to the FreedomCAR and Fuel Partnership. The applicability of each tool is summarized in Figure 2-1. Some of them have been developed, at least in part, by the Partnership and some have been developed outside. This set of systems analysis tools is now providing useful technical capabilities at the vehicle and fuel system levels. Increasingly, it is also providing capabilities at the implementation, impact, and policy levels.

The Phase 1 report reviewed the status of the available analysis tools and how they were then being used. The committee that wrote that report made recommendations in three areas:

- Continuing broad assessments to guide future research priorities and national transportation energy policy.
- Developing and using models to better understand consumer behavior during market transitions to new vehicle technologies and fuels.

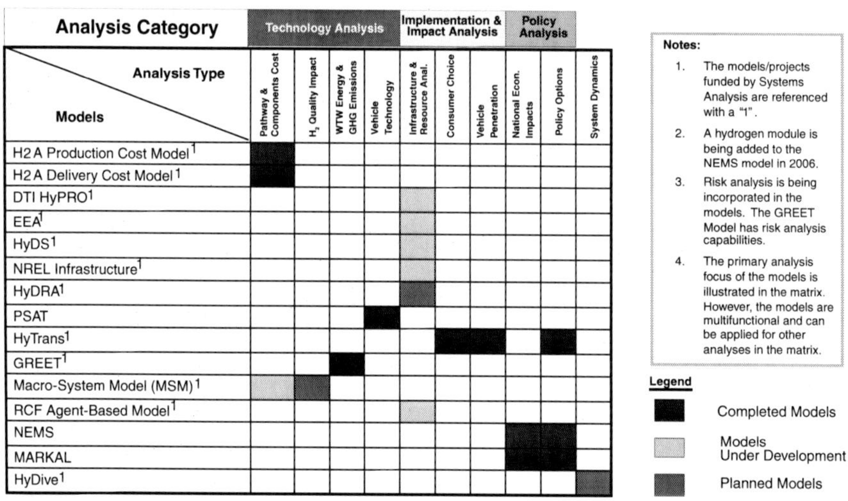

FIGURE 2-1 Models and analysis type matrix. DTI, Directed Technologies, Inc.; EEA, Energy & Environmental Analysis, Inc.; GREET, Greenhouse Gases, Regulated Emissions and Energy Use in Transportation; HyDive, Hydrogen Dynamic Infrastructure and Vehicle Evolution; HyDRA, Hydrogen Demand and Resource Analysis; HyDS, Hydrogen Deployment System; HyTrans, Hydrogen Transition; H2A, Hydrogen Technology Analysis; MARKAL, Market Analysis; NEMS, National Energy Modeling System; NREL, National Renewable Energy Laboratory; PSAT, Powertrain Systems Analysis Toolkit. SOURCE: F. Joseck and L. Slezak, DOE, Systems Analysis Effort, Presentation to the committee on April 25, 2007.

- Optimizing the systems analysis capabilities for the program management process.

Setting Goals, Targets, and Budgets

The overall goals and organizational structure of the Partnership were described in the introduction to this report. Detailed goals, milestones, and responsibilities can be found in the Partnership Plan (DOE, 2006). As was noted in the plan, the FreedomCAR and Fuel Partnership is not a legal entity, and the so-called "partners" do not have the responsibilities or rights of legal partners. Rather, the words "partnership" and "partners" are used in an informal sense to denote participants working together toward the stated goals of the group. Consequently, funding of the various activities that support the Partnership Plan goals and targets comes from the individual partners—namely, government and industry. The federal government contribution is provided by DOE and largely managed by the Office of Energy Efficiency and Renewable Energy (EERE) through its FreedomCAR and Vehicle Technologies (FCVT) program and its Hydrogen, Fuel Cells and Infrastructure Technologies (HFCIT) program, which have unique as well as shared responsibilities for various program activities that support Partnership goals (see Appendix A for the EERE organization chart). The individual companies fund activities that support not only overall Partnership goals but also their business and company requirements for commercializing the technology. The decision when and what to commercialize is left entirely to the individual company partners and is of necessity not transparent to the public owing to the competitive nature of the industry.

DOE supports the Partnership goals and targets via the DOE budget process. DOE has done a commendable job of gathering inputs from FCVT, HFCIT, and other DOE headquarters offices and laboratories and of creating an overall program plan, budget, and schedule that are aligned with the Partnership's goals and targets, including cost. This bottom-up approach is the first step in the federal budget process, and approval through the congressional appropriations process then allows DOE to fund activities designed to meet the technical and cost goals and milestones of the Partnership. New initiatives may arise as a consequence of high-level directives or as a result of technical advances or roadblocks. The Partnership is able to accommodate these efficiently since the DOE participants contribute to defining these initiatives as part of the overall budget process. As pointed out in the Phase 1 report, congressionally directed funding, when it occurs, has been detrimental in that it diverts funds from activities that are critical to the goals and milestones of the Partnership (NRC, 2005).

The original structure of the Partnership was appropriate given the state of the technology at the time. A broadly based program was established that supported both nearer- and longer-term approaches to reducing petroleum consumption and greenhouse gas (GHG) emissions. Hydrogen and fuel cell technology were the main, though not the sole, foci of the Partnership. Since its inception, good progress has been made, much learning has taken place, and new opportunities have become apparent (see Chapters 3 and 4). Now, it is appropriate to reexamine the Partnership's overall structure and balance. As noted above, systems analysis and simulation tools have been developed that can now be used to facilitate the planning process.

Program Assessment Methodology

The members of the FreedomCAR and Fuel Partnership should be commended for the progress that they have made in developing modeling tools and in beginning to apply them to various program elements. To date, this capability has lagged the overall technical program development and has been underutilized—that is, used for the most part only when requested by the various technical teams. There is no lack of technical review of the individual program elements, but what is missing is analysis of their quantitative impact on the overall goals of reducing petroleum use and air pollutant and greenhouse gas emissions. Tools for estimating this are being worked on; one example is the Macro System Model (MSM), which is scheduled for completion in 2008. This capability is critical to gaining a realistic assessment of the impact of various technology options and should be used extensively to validate or modify the program structure. A recent series of reports by the NRC Committee on Prospective Benefits of DOE's Energy Efficiency and Fossil Energy R&D Programs delineates an approach that employs decision analysis, panels of independent experts, technical risk and market risk, and takes into account DOE's role in technology development vis-à-vis that of the private sector. This approach is one way to make more realistic assessments of the potential impacts of technology development programs (NRC, 2005; 2007). Independent review of the analyses undertaken by DOE is critical to developing credible estimates of potential benefits. Such independent external panels could provide input to program managers and strategic planners.

Recommendation. DOE should accelerate the development and validation of modeling tools that can be used to assess the roles of various propulsion system and vehicle technologies and fuels, and utilize them to determine the impact of the various opportunities on the overall Partnership goals of reducing petroleum use and air pollutant and greenhouse gas emissions. Sensitivity analysis, from worst case to optimistic scenarios, should be performed to assess these impacts. Models, input data, and assumptions should be independently reviewed in order to validate and refine the models and lend credibility to the conclusions derived from them.

Program Management

A program of this scope requires effective decision-making and strategic planning processes to ensure that its objectives are appropriately focused and that good progress is being made. Program management requires a variety of system analysis tools to accomplish this, and during the 14 years of the two programs, several such tools have been developed (see Figure 2-1). The Partnership's activities need well-structured technical systems analysis assessments to provide detailed program management guidance. Increasingly, the vehicle performance model Powertrain Systems Analysis Toolkit (PSAT) and vehicle cost models are being used to do this. In fact, the continuing development of the PSAT, now centered at DOE's Argonne National Laboratory, has brought it to state-of-the-art status, and PSAT is used increasingly within the automotive industry and in R&D organizations. While several technology trade-off studies have been carried out, and such Partnership activity is continuing, in the committee's judgment these systems analysis models need to be used in a more ongoing, more structured way, as an essential component of overall program management. This type of major component and overall technical and cost analysis needs to be better integrated with the activities of the various technical teams to assess progress, review targets and goals, evaluate trade-offs, set priorities, and provide the information for go/no-go decisions. As noted in the preceding recommendation, independent review and validation of models is critical to the credibility of the conclusions derived from them.

Recommendation. The FreedomCAR and Fuel Partnership should use its technical and cost systems analysis capabilities as an essential component in program management to assess progress in meeting technical and cost targets, to examine the impact of not meeting those targets, to adjust program priorities, and to make go/no-go decisions.

Strategic Planning

The original FreedomCAR and Fuel Partnership started with a presidential commitment to request $1.7 billion over 5 years (FY04 to FY08). An important question for the Partnership's Executive Steering Group is how the Partnership should evolve as it continues to address the long-term goals related to reducing petroleum consumption, improving the nation's energy security, and achieving large reductions in greenhouse gas emissions. The Partnership's current implementation plan for key program areas, especially the development of fuel cells and hydrogen production and distribution technologies, already has key decision points between now and 2015, when it is anticipated that decisions about the commercialization of hydrogen-based technologies should start to be made. Thus a continuation of the Partnership is anticipated. Given the rising importance now being given to the transportation sector's energy consumption and environmental impacts, the committee believes it is appropriate, even essential, for the Executive Steering Committee to develop a strategic plan for the next phase of our nation's transportation technology development efforts.

In the committee's judgment, responding effectively to the two key issues the U.S. transportation system is facing—energy security and reducing greenhouse gas emissions—will require a wide range of new and improved propulsion system and vehicle technologies and new fuels. This future mix may well include fuel cells with hydrogen as a major energy carrier. It may also connect transportation to our electricity grid through the development of advanced battery technology for use in plug-in hybrids. It will also include much improved versions of our current ICEs and transmissions, substantial vehicle weight reduction and some vehicle size reduction, in conventionally configured vehicles and in hybrids. Biofuels will be contributing also, as will liquid transportation fuels obtained from oil sands, heavy oil, oil shale, coal, and natural gas. These new and improved vehicle and fuels technologies will require substantial coordinated R&D to efficiently prepare them for commercial deployment. Some will require basic and applied science research to overcome technology hurdles; others will require innovations to allow using these various forms of energy in transportation much more effectively.

It is, therefore, timely and important that the Partnership's leadership develop a strategic plan for the next phase and a long-term vision of the nation's collaborative vehicle and fuels technology R&D program, with appropriate industry partners. Given the importance of the energy issues the U.S. transportation system now faces, the Executive Steering Group should examine critically and broadly all potentially promising nearer- and longer-term options for reducing petroleum and energy consumption and greenhouse gas emissions in order to develop a balanced program to achieve the short-term goals and long-term vision. This strategic review should be done in the context of other ongoing domestic and international activities in vehicle and fuel technologies.

Recommendation. The FreedomCAR and Fuel Partnership's Executive Steering Group should establish a high-level planning group to develop a strategic plan appropriate for the next phase of the nation's collaborative vehicle and fuels technology R&D program.

Alternatives to Hydrogen

Since the committee's Phase 1 evaluation of the FreedomCAR and Fuel Partnership, interest in vehicle propulsion system options that could use energy from the electric grid has risen sharply. The primary technology—the plug-in hybrid electric vehicle (PHEV)—would use advanced batteries with substantial energy storage capacity so that the vehicle would have an all-electric range of 20-40 miles or otherwise use wall-plug electricity to replace much of the onboard fuel consumption. It should be noted that for a battery of sufficient size, recharging times are generally long; thus recharging is often thought of as occurring overnight. A benefit of this is better power plant utilization and lower electricity cost. This concept would solve the limited range of all-electric vehicles by incorporating a gasoline ICE and a generator so that as the battery runs down the engine can power the vehicle and the battery can be recharged by the engine and regeneratively by braking. The maximum engine power and electric motor power in such a configuration are usually comparable in magnitude.

This technology offers a way to share the energy used in transportation between a liquid fuel and electricity. The PHEV concepts currently being proposed roughly halve a vehicle's consumption of liquid fuel (gasoline, diesel, or a biofuel such as ethanol) per mile (Kromer and Heywood, 2007). Once it is widely deployed and used, this technology would significantly reduce the U.S. light-duty vehicle fleet's consumption of petroleum-derived fuels (gasoline and diesel).

The overall energy consumption per mile for a PHEV with electricity produced mainly from coal, natural gas, and nuclear energy and for that of a gasoline-fueled HEV are comparable, since engine efficiencies and system-average electricity-generating efficiencies are comparable (Kromer and Heywood, 2007). The GHG emission impacts are also comparable unless low-carbon-emitting electricity generation technologies are utilized. Given that the U.S. electric grid system is projected to continue to be dominated by coal- and natural-gas-fired power plants, reduced carbon emissions would require effective carbon capture and sequestration technologies to be widespread.

The major challenges to the successful development of PHEVs are battery performance and durability along with overall cost. Also, assessing the impact of a growing demand on the electricity grid by the transportation sector and for recharging capabilities is in its early stages. The Partnership has initiated programs directed toward PHEV development: advanced batteries, power electronics, electric motors, and systems simulation and testing in FY07 and FY08. About 40 percent of the HEV funding is being allocated to these PHEV technologies, with about two-thirds of that focused on advanced batteries. These activities are reviewed more fully later in this report.

It is now apparent that PHEVs running on electricity and liquid hydrocarbons (which might be augmented by biofuels) is a plausible parallel approach to fuel cell propulsion system technology and hydrogen for achieving major reductions in U.S. petroleum use and GHG emissions. This was not clear at the time the committee last evaluated the FreedomCAR and Fuel Partnership during Phase 1. Because PHEVs can be considered as both complementing and competing with hydrogen fuel cell vehicles, any comparison between PHEVs and hydrogen fuel cell vehicles must be made with the same initial assumptions. In order for the fuel cell vehicle to meet the goal of a 300-mile range, vehicle structural weight must be reduced by 50 percent (see discussion in Chapter 3, section Structural Materials). The same vehicle structural weight must be assumed when comparing the cost or performance of PHEVs and fuel cell vehicles. Accordingly, the

Partnership should consider more fully how best to pursue this parallel technology and fuel.

Recommendation. The Partnership management should assess how best to pursue PHEV technology within the FreedomCAR and Fuel Partnership program and determine the cost and performance merits relative to hydrogen fuel cell vehicles using the same vehicle structural weight for both systems.

Role of the Federal Government and Industry

The Partnership's ultimate goal is to reduce the dependence of the nation's personal transportation system on imported oil and minimize harmful vehicle emissions, without sacrificing freedom of mobility and freedom of vehicle choice (DOE, 2006). Meeting the technical and cost targets of the largely DOE-funded program does not guarantee that technologies will be adopted by the automobile manufacturers and energy companies. Because all of these activities are precompetitive and the industrial partners have yet to commercialize technology from the Partnership, it is difficult to assess its ultimate true impact, which may not be realized in the broadest sense for many decades. However, the mere fact that the industry is actively engaged in setting goals and targets suggests that its needs are being addressed and there is good potential for technology transfer. The committee that wrote the Phase 1 report recommended (Recommendation 2-14, Appendix D) that the Partnership and USCAR leadership assess the process for technology transfer and make it as effective as possible. While DOE has done a good job of pursuing this objective by promoting technology transfer mechanisms and workshops, such transfer is ultimately the responsibility and choice of industry. The committee understands that these decisions are made in a closed and competitive environment, but industry should cooperate with the DOE as much as possible to establish a database of technology transfer case histories, including those of the earlier PNGV program. This would serve to provide useful models for improving this and any potential future government/industry partnerships.

An important role of the federal government is to invest in high-risk, high-payoff activities that are unlikely to be supported by industry. The committee commends DOE for requesting increased support for the Office of Basic Energy Sciences (BES) program and for including the program in its Annual Hydrogen Merit Review in 2007. The BES program has been responsive to the FreedomCAR and Fuel Partnership by focusing on critical issues and materials that are the building blocks of technology platforms envisioned by the Partnership. It is here, in the BES program, where a major breakthrough might occur that could dramatically alter the course or outcome of the Partnership.

FreedomCAR and Fuel Partnership in the Marketplace and Policy Context

It is industry's responsibility to commercialize the technologies that will be needed to achieve the goals of the FreedomCAR and Fuel Partnership. If there is not a sound business case to do so, it is unlikely that this will occur even if technical and cost targets are met. There are many factors, other than the current state of technology development, that could influence whether or not these technologies are introduced and in what time frame. Market interventions such as cap-and-trade programs, motor fuel taxes, corporate average fuel economy (CAFE) standards, fees and rebates imposed at time of vehicle purchase, and subsidies for specific technologies and fuels could influence the introduction of new technologies and fuels, especially the transition to a hydrogen economy and its timing. These were discussed in detail in the Phase 1 report. Recommendation 2-15 (Appendix D) asked DOE to analyze the implications of alternative market interventions for the technical goals of the Partnership and to use that analysis in its policy deliberations. In its response of April 2, 2007, to the recommendations in the Phase 1 report, DOE indicated that a joint draft report is being developed with the Environmental Protection Agency (EPA) and the Department of Transportation (DOT), Analysis of Market-Based Approaches for Reducing Fuel Consumption, which assesses the benefits from various policy approaches (DOE, 2007, p. 45). While DOE has claimed that this recommendation is not the responsibility of the Partnership, the committee concluded that such policy actions could have a profound impact on program structure and balance and should be included in program planning.

To support an assessment of market interventions, a better understanding of the expected market response to new vehicles and fuels at significantly different prices and with significantly different performance and operating characteristics is important. Such modeling and assessment is a challenging task. However, analytical approaches have been proposed by Cook and others that estimate economic value based on vehicle attributes, and these have been shown to work in a number of automotive cases (Cook, 1997; Donndelinger and Cook, 1997). Trade-offs between range and useful passenger space with vehicle value can be estimated with such techniques. The Oak Ridge National Laboratory effort by David Greene and others that builds on Cook's work could be developed to assist in this task (Greene et al., 2004). The committee believes that initiating efforts in this market response area is important and would prove fruitful.

Recommendation. DOE should utilize its modeling capability to assess the impact of market interventions on both the technical goals of the FreedomCAR and Fuel Partnership, and their overall potential impact, and use these assessments to inform the R&D planning process.

Recommendation. The Partnership should evaluate the potential for analyzing and predicting market responses to the vehicle technologies and fuels that may result from Partnership efforts to better inform its assessments of the new technologies that are likely to be needed to meet the nation's goal of reducing petroleum consumption and greenhouse gases.

SAFETY

Overview

While an exemplary hydrogen safety record will not ensure the success of the fuel cell and hydrogen technologies under development by the Partnership and the eventual transition to a hydrogen economy, a poor safety record may delay or inhibit the widespread use of hydrogen. The goals and objectives of the broad safety portion of the Partnership are to develop practices and procedures that will ensure safety in the operation, handling, and use of hydrogen and hydrogen systems for all DOE-funded projects and to implement these practices and lessons learned to promote the safe use of hydrogen.

The goals and objectives of the narrower codes and standards portion of the program are as follows:

- To perform the underlying research to enable codes and standards to be developed for the safe use of hydrogen in all applications and
- To facilitate the development and harmonization of domestic and international codes and standards.

Activities under the umbrella "safety, codes and standards (SCS)" have been funded at $4 million to $5 million or so per year over the last few years—well below requested levels. In FY07, SCS received a large increase, to $13.8 million. The codes and standards portion is planned and overseen by the Partnership's codes and standards technical team (see Figure 1-1). The safety part is administered by DOE headquarters. DOT now has hydrogen safety resources in its own budget ($1.4 million actual in FY07 and $1.4 million requested in FY08).

The codes and standards portion of the Partnership, which includes the R&D Roadmap and National Template, aims to gain the support of the many organizations developing vehicle- and component-level safety standards. There is work on fueling station design tools and hydrogen quality standards. There is also work on fast fueling at up to 70 MPa (10,000 psi).

"Safety" consists of more than just following a set of codes and standards. It is quantitative and includes system design and methods of mitigating risk. Every component of a system may meet an appropriate code or standard, yet there can still be failures due to external events and system issues.

The safety part of the program includes a hydrogen safety panel, Web-based incident reporting and bibliographic databases; a best practices Web site is under development. There is also an extensive program on unintentional releases of hydrogen and on hydrogen behavior, safety sensors, and compatibility of other materials with hydrogen.

Responses to Safety Recommendations 2-5 to 2-8 in Phase 1 Report

The recommendations discussed in this subsection come from the Phase 1 report (NRC, 2005) and may be found in Appendix D of this report.

- *Recommendation 2-5: Formation of a crosscutting safety technical team.* The codes and standards technical team decided not to accept this scope and organizational recommendation. The committee's suggested step forward is covered in the first paragraph of the next section.
- *Recommendation 2-6: Vehicle standards and the National Highway Traffic Safety Administration (NHTSA).* The NHTSA vehicle safety program was delayed due to funding constraints but is now under way. It is not clear whether NHTSA is fully integrated into the codes and standards technical team since none of its milestones were shown on the roadmap. It is also not clear to the committee whether NHTSA is taking advantage of the hydrogen behavior work being conducted at Sandia, Livermore.
- *Recommendation 2-7: Publication, Openness, and Safety Documents.* The incident reporting system and bibliography have been implemented and are being well maintained. The best practices document is under development. NHTSA has special crash investigation teams that could, if so assigned, respond to accidents involving vehicles fueled by compressed natural gas or hydrogen.
- *Recommendation 2-8: Budget and schedule.* The FY07 DOE SCS budget increase is a good step forward. The SCS milestone chart should be extended to 2015. (See discussion in the next section.)

Discussion and Recommendations for the Phase 2 Review

The codes and standards technical team did not accept the Phase 1 recommendation that DOE should form a new technical team to cover all end-to-end safety aspects as well as codes and standards. That is their prerogative. However, the overall safety aspects are still important, and this gap must be filled by DOE, which should assign an organization to head this assignment. The assignment could be viewed as an extension of the existing quantitative risk analysis task, which is currently focused on filling stations and should be adequately funded and extended. The priority for expansion should be (1) the vehicle and (2) the fuel distribution system. Since the United States already produces approximately 25 percent of the hydrogen needed for the conversion of the vehicle fleet to hydrogen, the safety analysis of the production portion of the end-to-end system is the lowest priority (because production has an adequate track record). Onboard liquid hydrogen storage and home refueling should also be included.

This analysis should be at a high level and not overly detailed. It could affect some of the technology and pathway choices. Further depth in the risk analysis can be delayed until 2010-2015.

The committee notes that there was a substantial increase in the SCS budget beginning in FY07 (to $13.8 million). For the first time the actual budget was near the requested budget. It is important that the SCS budget remain adequate and stable for the coming years. DOE should protect this funding to ensure that the milestones are met.

The DOT part of the program is just getting started, is behind schedule, and is grossly underfunded ($1.4 million per year). NHTSA's Four-Year Plan for Hydrogen, Fuel Cell, and Alternative Fuel Vehicle Safety Research was based on expenditures of $4 million to $5 million per year. The Research and Innovative Technology Administration (RITA) and the Pipeline and Hazardous Materials Safety Administration (PHMSA) and other parts of DOT also have important hydrogen safety roles and need to be adequately funded. DOT needs to develop a long-range plan with budget estimates and milestones to 2015. It should be comprehensive and include all relevant administrations and agencies within DOT. These milestones should be integrated into the codes and standards technical team roadmap.

It is doubtful that the SCS milestones are consistent with the progress to date and with the delays in getting full funding for the program. The current milestones only extend to 2010, while the rest of the program has been using a 2015 planning horizon. Clearly, a lot of safety-related work will have to be done from 2010 to 2015.

While DOT is on the technical team, the team's milestone chart does not include DOT's milestones. NHTSA recently initiated its R&D program, and RITA has done an extensive gap analysis for pipelines, distribution trucks, and other DOT regulatory areas. Perhaps RITA and PHMSA should also be included on the technical team.

It is unlikely that the necessary SCS work will be completed by 2010. Realistic schedules should be adopted, new work identified, and milestones planned out to 2015, as has been done for other program elements. DOT milestones and its program should be integrated into the milestones and roadmap of the codes and standards technical team. When developing milestones, it is important to consider that developing and finalizing new federal regulations takes 7-10 years (or even longer), and lack of regulation could significantly impede the introduction of hydrogen vehicles into the market. Additional funding for more detailed work should be planned for the out-years. The details of the out-year work will of course depend on what is learned over the next few years. Continuing work is likely to be needed in hydrogen quality, sensors, and risk analysis. Real-world experience should also be factored in and revisions made to various codes and standards, if needed.

Possible areas of new work include hydrogen vehicle sensors (and associated standards); thermally activated pressure relief devices (PRDs), which are sensitive to a line or area (rather than to a point); development of a localized fire test and a full-scale vehicle burn test; research on whether to allow insulation to provide fire protection for a short time; special safety considerations for reactive metals such as storage hydrides and structural magnesium; and development of a hydrogen compatibility test at the component or assembly level.

It appears that the introduction of some kinds of high-energy lithium ion batteries has been delayed, in part owing to safety concerns, especially under abusive conditions. Since such batteries would benefit the performance of HEVs, PHEVs, and pure EVs as well as hydrogen/fuel cell hybrids, it is important to pay even more attention to battery safety issues, including subsystem- and system-level approaches to protection.

The creation of two databases, one on incidents involving hydrogen and one for a hydrogen bibliography, will be useful in promoting safety. The committee encourages DOE to continue to develop, publish, and update the best practices document.

Many of the hydrogen vehicle safety components and subsystems (and associated codes and standards) have evolved from analogous components used for compressed natural gas (CNG) vehicles. It is important to collect and analyze CNG safety experience while it is still available. This should include filling station incidents, vehicle failures attributable to such things as tank leaks, tank bursts, PRD failures (failure to open in a fire, as well as inadvertent opening without a fire), and other component failures or leaks.

DOE should establish a program to collect and analyze failure data on CNG and hydrogen components, subsystems, vehicles, and fueling stations. These data should be statistically analyzed to assess field reliability and to determine if this level of reliability will be adequate if the majority of U.S. vehicles are fueled with hydrogen. NHTSA data should also be included and analyzed.

The committee heard a briefing on hydrogen compatibility (including embrittlement), which is being worked on by Sandia National Laboratories. This is an important area and should be extended to nonstructural materials in the future. DOE should convene a review by outside experts and hydrogen material users to examine the scope of the various materials to be tested, the priority for testing, and the test procedures and conditions. The prioritization should carefully consider the likelihood that a material will be used and not try to cover every last material. At least some of the reviewers should be from the academic community.

The Sandia National Laboratories work on unintended hydrogen release has shown great progress with unignited and ignited jets. New work on releases with delayed ignition is just beginning. Such releases can result in explosions with damaging overpressures. It is understood that Sandia will focus on open air releases with and without barriers; the National Renewable Energy Laboratory (NREL) on residential garages; and the National Institute for Standards and Technology (NIST) on commercial garages. It is not clear whether some organization has been assigned to study releases in tunnels.

The current program has a small program element called "parking certification." This task is currently limited to hydrogen production from the charging of lead acid batteries. This task should be expanded to include the general topic of parking hydrogen vehicles in buildings (such as residential and commercial parking garages, and repair facilities). Small, medium, and large leaks (such as PRD activation) should be addressed.

DOE should accelerate this unintended release and parking structure work so that the data it collects can feed into updates of National Fire Protection Association (NFPA), International Code Council (ICC), and other codes and standards. This would also support the work on fueling station footprints (separation distances). Station design parameters (such as reduced on-site storage, reformer turn-down ratios, more frequent delivery) might also allow a smaller station footprint. It would also be useful to document conditions under which hydrogen jets could self-ignite.

Four large demonstrations involving the automotive and energy companies are conducted under the technical validation part of the Partnership. For proprietary reasons, the safety plans for these demonstrations are confidential. Safety incident information is also confidential, and only summaries are released. It is important that there be an independent review of these safety plans and field incidents, and that the safety lessons learned be shared with the hydrogen community. Relevant incidents should be added to the incident database.

Appropriate Federal Role

Work on SCS is an essential federal role. The individual companies and states cannot do it on their own. The manufacturers want and need uniform national (and, hopefully, international) standards so they can market worldwide. The various regulatory agencies (which reside mainly in DOT) need to be adequately funded in order to develop scientifically based standards in a timely fashion. They can then work with international groups to harmonize the standards on a global basis. This is inherently a government function.

Recommendations

Recommendation. DOE should establish a program to address all end-to-end safety aspects as well as codes and standards. Such a program could be viewed as an extension of the current quantitative risk analysis activity, which is focused on the filling station. This task should be adequately funded and expanded. The priority for expansion should go to (1) the vehicle and (2) the fuel distribution system.

Recommendation. The Department of Transportation (DOT), including all relevant entities within DOT, should develop a long-range, comprehensive hydrogen safety plan with budget estimates and milestones to 2015. The milestones developed in this plan should be integrated into the codes and standards technical team milestones and roadmap.

Recommendation. The codes and standards technical team should extend the planning horizon in its plan to 2015, integrate the DOT milestones into its own milestones and roadmap, and make the safety and codes and standards milestones consistent with funding levels and progress to date.

Recommendation. DOE should establish a program to collect and analyze failure data and field experience including data from the National Highway Traffic Safety Administration on compressed natural gas (CNG) and hydrogen components, subsystems, vehicles, and fueling stations.

Recommendation. DOE should convene a review by a panel of independent outside experts of the hydrogen compatibility of materials, prioritize the materials to be tested, taking into account the likelihood of their application, and review test procedures and conditions.

Recommendation. DOE should accelerate work on delayed ignition of unintended hydrogen releases, including in parking structures and tunnels, in support of various efforts to develop and revise building codes.

TECHNICAL VALIDATION

Technical validation activities, also referred to as learning demonstration programs, are very important for validating current component and systems concepts and uncovering previously unknown issues. They are establishing many systems and component engineering parameters for a complete operating hydrogen supply and fuel cell transportation system. In general, the committee notes that

- These cooperative programs are well designed,
- Information is being collected from both vehicle and infrastructure components, pooled, and shared,
- This program is helping to guide the technical teams as well as the systems and modeling efforts and helping to establish appropriate program priorities,

- There are some indications of a lack of adequate DOE support and balance for the vehicle and infrastructure side of these programs.

The FreedomCAR and Fuel Partnership includes a variety of R&D and technical validation activities for fuel cell vehicles and hydrogen fuel systems. Just under 10 percent of the total FY07 budget for the Partnership is focused on validation activities (however, this drops to under 7 percent in the FY08 budget request), and these are providing extremely valuable information for the overall program. Any advanced technology, no matter how well tested by its developers, will show unanticipated characteristics when placed in the hands of the users. Moreover, all complex technologies require at least alpha and beta prototype versions before a reasonably reliable product is possible. Feedback from actual use is especially important for the Partnership because of the long-term, high-risk research agenda and because public safety must be ensured in the face of highly energetic materials—for example, hydrogen and high-voltage batteries.

The committee applauds the Partnership for managing an effective technical validation program, as was recommended in the Phase 1 report. The committee is particularly pleased at the effective communication of findings from the technical validation program via the Web to all Partnership elements and, more broadly, to forums where the public can obtain information as well. In line with the committee's earlier recommendations, these technical validation efforts have appropriately emphasized safety and communication to first responders. The integration of the learning demonstration findings with the systems analysis effort is also appropriate and in agreement with the Phase 1 report recommendations. Thus the learning demonstration program should continue to be considered an essential component of the Partnership. Rather than attempting to demonstrate that the technologies are commercially ready, the Partnership should continue to collect and analyze the experience of the early adopters of hydrogen vehicles and fuels infrastructure technologies in order to inform the various research programs. Further, having private companies as partners will help disseminate the learning beyond DOE. Recognizing that the learning demonstration program is now in an active stage, the committee recommends several tasks for the DOE to consider as this program unfolds.

Recommendation. DOE should continue to disseminate the results of the technical validation activity to supporting organizations outside the Partnership in order to promote widespread innovation and competition. DOE management needs to systematically evaluate the information being generated by each project to determine when the project should be terminated based on its relevance and on the value of the information. On the other hand, DOE management should not prematurely drop support for the overall technical validation and learning demonstrations as their importance cannot be overemphasized. DOE and the Partnership should develop a long-range plan for technology validation that continues until at least 2015.

Recommendation. DOE management should maintain adequate support for technical validation as it is essential to the overall Partnership. This support should be balanced and cover both the vehicles themselves and the fuel infrastructure needed. To achieve the rapid learning that the overall project requires, DOE should also keep the validation activities focused on their primary purpose—the accumulation, analysis, and dissemination of experience from the field. Safety should be stressed throughout the learning demonstration program, because an accident early on could attract publicity out of proportion to its true consequences.

BUILDING PARTNERSHIPS WITH NEW VENTURES

Start-up ventures have often provided an essential stimulus for large-scale technology transitions. This holds true whether the new venture acts alone or in partnership with established companies. In the latter case, the new venture can bring a fresh approach to the partnership—a technology outside the scope of the incumbent or an innovative business model—while the mainstream industry provides investment capital and channels to markets.

Consider the commercial introduction of the personal computer (PC), for example. To move the PC from a hobbyist's toy into the mainstream market, the early competitors, chiefly Apple and IBM, needed some application that offered compelling value to business users. That application proved to be VisiCalc, the first practical electronic spreadsheet. Despite their capabilities in microelectronics, neither Apple nor IBM was able to conceive such an application. Instead, it came from a start-up Boston company, Software Arts (later renamed VisiCorp). VisiCalc became the key application offered in the Apple II and the IBM PC, and by 1981 the mainstream PC market was fully launched.

Similarly, much evidence suggests that entrepreneurs interested in automotive innovation will respond vigorously to clear signals about any potential opportunity. Consider hydrogen technology markets, for example, which have already seen a vigorous response to government signals. When the FreedomCAR program was announced, virtually all hydrogen-related companies saw increases in their share prices. And from 1998 to 2001, signals from the auto industry about imminent introduction of hydrogen-fueled vehicles led to a surge in private and public capital into entrepreneurial companies that offered technologies able to serve this anticipated new market.

DOE seeks to stimulate new, technology-based ventures through its Small Business Innovation Research (SBIR) and Small Business Technology Transfer (STTR) programs. These have been reviewed in depth elsewhere,[1] and here we focus only on those aspects relevant to the FreedomCAR and Fuel Partnership.

---

[1]See NRC, *An Assessment of the Small Business Innovation Research Program at the Department of Energy*, Washington, D.C.: The National Academies Press, forthcoming.

Supported at the level of 2.5 percent and 0.3 percent of the DOE research budget, respectively, the SBIR and STTR programs offer significant resources to start-up technology companies—over $100 million per year, DOE-wide, in recent years. This investment is allocated among the DOE programs in approximate proportion to their research budgets. The overall program is administered by the DOE's Office of Science (SC).[2]

Much evidence in the form of case studies suggests that the SBIR program can indeed accelerate innovation in ways that serve the goals of the Partnership. These individual case studies are, of course, helpful and should be continued. But by themselves, they cannot provide the kind of systematic insight that leads to improvements in program management because they look only to success and not to the sources of failure. Additional examples of the kinds of management-relevant questions that could be addressed include these:

- How far in advance of the annual solicitation should the technical topics be announced?
- How often should these topics, which are currently issued on an annual basis, be changed?

Thus, the committee believes that the SBIR/STTR program could make an even greater contribution by allowing a more complete understanding of how the management of the program influences the success rate of the funded companies.

Further, contact with the new venture community appears largely limited to the SBIR/STTR programs and contracts with Partnership members or national laboratories. This limited contact might allow the Partnership to overlook many nascent ventures that could add significant value its mission. Expanding the range of contact between Partnership and the entrepreneurial community beyond the SBIR/STTR might capture some of this value. This has been done with apparent success in other DOE programs. Consider, for example, the Industry Growth Forum organized by NREL. This kind of forum could introduce entrepreneurs to the leading automotive and fuel companies and could also provide useful guidance on the most productive means for interaction. Additionally, DOE might consider some form of interaction with organizations that invest in first-stage ventures, both to increase participation in the SBIR/STTR program and to provide further channels for SBIR/STTR grantees. The committee makes two recommendations that would help to accomplish this.

[2]More information is available at <http://www.science.doe.gov/sbir/newweb/about_sbir.htm>.

Recommendations

Recommendation. DOE should conduct a systematic assessment of the success (or failure) of all its SBIR/STTR-funded companies rather than selected case studies.

Recommendation. The Partnership should seek ways beyond the SBIR and STTR programs to improve communications between it and the entrepreneurial community.

ENVIRONMENTAL ISSUES

The Phase 1 report noted the importance of understanding the environmental implications of the full fuel cycle, from source to end use, in a hydrogen economy. Hydrogen, like electricity, is an energy carrier and must be produced using a primary energy source. As discussed in Chapter 4, a number of approaches and primary energy sources could be used to produce hydrogen to fuel vehicles: the reformation of natural gas; the gasification of coal; high-temperature nuclear heat from advanced reactors to drive thermochemical processes; the gasification of biomass; or electricity from the grid or renewable energy sources (such as wind or solar) to electrolyze water to produce hydrogen.

Each of these approaches and technologies for producing hydrogen will have different land, water, and air impacts. They will also have different emissions of carbon dioxide (CO_2) and other GHGs, depending on the extent to which CO_2 sequestration is used as part of the fossil-fuel-based processes for producing hydrogen. In addition, as was also pointed out in the Phase 1 report, there could be impacts if hydrogen leaks into the atmosphere throughout the fuel cycle, ranging from impacts on the stratosphere to contributions to global warming (Ananthaswamy, 2003; Derwent, 2003, 2004; Tromp et al., 2003).

To understand the impacts across the full fuel cycle of producing, distributing, and using hydrogen, the Phase 1 report recommended that DOE, in collaboration with the EPA, should systematically identify and examine the possible long-term ecological and environmental effects of the large-scale use and production of hydrogen from various energy sources. These direct and indirect effects should include effects on land, water, and the atmosphere.

In its response dated April 2, 2007, to the recommendations in the Phase 1 report, DOE concurred with this recommendation (DOE, 2007, p. 23); in fact its SC is developing a fundamental understanding of the processes involved in biogeochemical cycling of atmospheric hydrogen. This knowledge will make it possible to perform a comprehensive assessment of the environmental impact of the release of hydrogen to the atmosphere from large-scale use and production. DOE will share the results of the assessment with the EPA and explore collaboration possibilities. The DOE response to the Phase 1 recommendation also contained the following (DOE, 2007, p. 23):

1. The DOE Hydrogen Program is planning a Programmatic Environmental Impact Statement (PEIS) in the 2011-2014 timeframe, culminating with a final report in 2014. Conducting the PEIS prior to then would be premature, since R&D is still in progress and the actual technologies to be employed during implementation of a hydrogen fuel infrastructure are not yet determined.

2. During the RD&D timeframe leading up to the PEIS, the Program will be conducting a Strategic Environmental Review (SER). A SER involves the collection and assessment of potential environmental issues arising from and reported by the various R&D projects funded by the Program. A database of these issues will be maintained, both (a) to be a source of information for the PEIS which follows and (b) to identify particular items which might require Program analysis during the R&D phase to better understand their potential impacts if the resulting technologies were implemented in the objective hydrogen infrastructure.

3. The NAS [Phase 1 report] also calls for a study of environmental effects of hydrogen technologies. An FY 2007 solicitation is commissioning a study to respond to this requirement.

4. In 2010 and 2011, a joint study with EPA is planned to complete a comprehensive examination of the effects of hydrogen on the atmosphere.

The committee will continue to pursue this subject in future reviews of the Partnership and monitor the progress that is being made on understanding the various environmental impacts that may be associated with the large-scale production and use of hydrogen.

REFERENCES

Ananthaswamy, A. 2003. Reality bites for the dream of a hydrogen economy. *New Scientist*, November 15.

Cook, H.E. 1997. *Product Management: Value, Quality, Cost, Price, Profit and Organization.* Amsterdam: Kluwer Academic (formerly Chapman & Hall).

Derwent, R. 2003. Climate implications of a hydrogen economy. Berkshire, England: Climate Research, The Meteorology Office. Available on the Web at <http://www.cambrensis.org/hydropresentation.htm>.

Derwent, R. 2004. Global warming consequences of a future hydrogen economy. *Issues in Environmental Science and Technology*. Vol. 20: *Transport and the Environment.* London, England: Royal Society of Chemistry.

Department of Energy (DOE). 2006. *Partnership Plan. FreedomCAR & Fuel Partnership.* Washington, D.C.: U.S. Department of Energy, Office of Energy Efficiency and Renewable Energy. Available on the Web at <http://www.eere.energy.gov/vehiclesandfuels/pdfs/program/fc_fuel_partnership_plan.pdf>.

DOE. 2007. *Actions and Evidence Report, April 2.* Submitted to the National Research Council Committee on Review of the Research Program of the FreedomCAR and Fuel Partnership, Phase 2, documenting responses to recommendations by DOE to the Phase 1 report by the National Research Council.

Donndelinger, J., and H.E. Cook. 1997. Methods for analyzing the value of automobiles. *SAE 1997 Transactions, Journal of Passenger Cars* 106:1263-1281.

Greene, D.L., K.G. Duleep and W. McManus. 2004. Future potential of hybrid and diesel powertrains in the U.S. light-duty vehicle market. ORNL/TM-004/181, Oak Ridge, Tenn.: Oak Ridge National Laboratory.

Kromer, M., and J.B. Heywood. 2007. *Electric Powertrains: Opportunities and Challenges in the U.S. Light-Duty Vehicle Fleet.* Report 2007-03-RP. Cambridge, Mass.: Massachusetts Institute of Technology, Laboratory for Energy and the Environment. Available on the Web at <http://lfee.mit.edu>.

National Research Council (NRC). 2005. *Prospective Evaluation of Applied Energy Research and Development at DOE (Phase 1): A First Look Forward.* Washington, D.C.: The National Academies Press.

NRC. 2007. *Prospective Evaluation of Applied Energy Research and Development at DOE (Phase 2).* Washington, D.C.: The National Academies Press.

Tromp, T.K., R.L. Shia, M. Allen, J.M. Eiler, and Y.L. Yung. 2003. Potential environmental impact of a hydrogen economy on the stratosphere. *Science* 300 (June 13):1740-1742.

3

Vehicle Subsystems

INTRODUCTION

At the time of the Phase 1 review of the FreedomCAR and Fuel Partnership, the Partnership had been under way for a relatively short time, although it did have the advantage that a number of the technologies under development were part of the Partnership for a New Generation of Vehicles (PNGV) program, which had been initiated in 1993. Since the Phase 1 report was issued, there have been significant changes in some external influences, such as a large increase in the price of gasoline and heightened interest in carbon dioxide (CO_2) contributions to global warming. These changes could make achieving the program goals even more beneficial in the long term and, consequently, more important to the nation. Since the program objectives are obviously in the national interest, and since most of the high-risk activities would not be undertaken without governmental support, DOE involvement is clearly justified.

The long-range goals of the Partnership—to transition to a transportation system that uses sustainable energy resources and produces minimal criteria or net carbon emissions on a life-cycle or well-to-wheels basis—are extremely ambitious. The difficulties are compounded when the additional constraints associated with the Partnership are imposed: energy freedom, environmental freedom, and vehicle freedom. These goals and associated constraints effectively eliminate the continued simple evolution of the gasoline-fueled internal combustion engine (ICE) vehicle as a possible answer. "Sustainable energy resources" and "energy freedom" both suggest non-petroleum-based alternative fuels. The emphasis on "net carbon emissions" and "environmental freedom" suggests that CO_2 and other emissions from the production and consumption of alternative fuels should be reduced, through highly efficient processes, to minimize adverse environmental effects. Finally, "vehicle freedom" implies that the fuel and onboard energy conversion systems should not limit the options and choice that buyers expect to have available in their personal vehicles. These goals, if attained, are likely to require new transportation fuel(s) utilized in more efficient power plants in lighter vehicles having reduced power requirements and equivalent utility and safety.

DOE envisions that the path to achieving the long-term goals of the FreedomCAR and Fuel Partnership involves improvements in ICEs, a transition from improved gasoline- and diesel-fueled ICE vehicles to a greater utilization of gasoline- and diesel-fueled hybrid electric vehicles (HEVs), the development and implementation of plug-in hybrid electric vehicles (PHEVs), more utilization of hydrogen-fueled ICEs and HEVs, and—finally—hydrogen-fueled fuel cell vehicles (FCVT, 2004). For this transition to take place, the industry will require enhanced technology in many areas so that it can develop new vehicle subsystems and vastly improved vehicles. The DOE-sponsored activities described in this section are intended to provide understanding that will enable the needed technologies to be successful. The scope of the technologies needed is broad and the timescales for implementation are short. Competitive pressures dictate that if a technology appears to be promising it will be rapidly integrated into industry's implementation plans. Consequently, continued close collaboration between DOE and industry is necessary to allow these technologies to transition into the industrial laboratories and development programs and then to identify new critical areas where enhanced understanding will be most beneficial.

An example of technology that has progressed from concept demonstration stage to implementation is exhaust gas aftertreatment from lean-burn engines. Various engine manufacturers and original equipment manufacturers (OEMs), both domestic and foreign, have or are planning to introduce lean oxides of nitrogen (NO_x) traps, selective catalytic reduction, and diesel particulate filters to their production models. Therefore, it is appropriate that DOE funding for these activities should be curtailed and redirected to areas where fundamental understandings are lacking. This observation was made in the Phase 1 report and recommended that the Partnership redirect its pursuit of novel emission control technologies and plan for, analyze, and seek solutions for emission problems associated with emerging fuels, fuel infrastructure, and propulsion systems (see Recommendation 3-3 in Appendix D).

In response to the recommendation, DOE initiated a solicitation (DE-PS26_07NT43103) to address (1) E85-optimized engines, (2) enabling technologies for fuels and lubricants, and (3) efficiency of clean combustion and fuels development. Since some of these technologies are in production (e.g., flex-fuel vehicles), it is important that the Partnership carefully coordinate with industry to maintain programs that contribute most to a new understanding of the physical, chemical, and thermal processes impacting performance of the engine, the fuel, and the aftertreatment system.

The most direct way to enable near-term reductions in fuel consumption and emissions is by improving ICEs. Specifically, better understanding of the ICE combustion process and how emissions are produced could both increase efficiency and decrease engine-out emissions. Higher thermal efficiency reduces the fuel needed to produce a given power output, and lower engine-out emissions will allow the use of a simpler, probably less expensive exhaust aftertreatment system. Such improvements in ICEs, which could be implemented quickly, would benefit both conventional vehicles and HEVs.

The fuel cell subsystem is an energy converter with the potential to be more efficient than an ICE. However, the only fuel cell systems currently appropriate for transportation systems use hydrogen as fuel. The hydrogen can be stored onboard the vehicle in pure form or it can be extracted from hydrocarbon fuels and water using an onboard fuel processor. However, DOE effectively eliminated the latter alternative from its R&D portfolio after years of research determined that there was little prospect of meeting essential cost and performance targets within the program time frames. This means that sufficient pure hydrogen must be carried onboard the vehicle to meet range requirements a very challenging task given the space and weight limits of typical light-duty vehicles. This, in turn, places a high premium on reducing the mass of the vehicle and maximizing the efficiency of the energy converter.

Current experimental hydrogen-fueled fuel cell systems demonstrate efficiencies approaching 50 percent over a fairly wide range of operation. Further, such systems produce zero criteria emissions (occasional discharges of small quantities of hydrogen may occur). FreedomCAR and Fuel Partnership programs are addressing the performance, durability, and cost issues that need to be resolved so that fuel cells can become a viable option for personal transportation vehicles.

HEVs require compact, efficient, and low-cost power electronics and energy storage systems as well as other advanced electrical components to make vehicle costs and weights competitive with conventional vehicles. Many of these same technologies are applicable to fuel cell vehicles and PHEVs. Consequently, advances in the power electronics and electrical subsystems are critical for improved viability of both the midterm as well as the longer-term vehicles envisioned by the Partnership.

It is possible for HEVs and fuel cell vehicles to reduce fuel consumption by capturing some of the vehicle kinetic energy during deceleration and stopping. This requires some form of energy storage capable of accepting this energy and returning it to the drive train for propulsive power (called regenerative braking). The most likely form of such energy storage is electrochemical (batteries), but ultracapacitors are also being investigated. For such relatively small-scale energy storage, the most important parameters are cost per kilowatt ($/kW), specific power (kW/kg), cycle life, and calendar life.

With increased electric energy storage onboard, the vehicle could run for a significant distance without using the fuel cell or engine. This would add design flexibility to HEVs and reduce some of the performance requirements for fuel cells (e.g., start-up time and power ramp-up rate). Further increases in onboard energy storage capacity could enable PHEVs (vehicles whose batteries could be recharged by plugging them into a source of electricity while it is parked) and even all-electric vehicles. Both plug-in hybrids and all-electric vehicles would shift some transportation energy demand from petroleum-based fuels to the electric grid, which is mostly non-petroleum-based but not emissions free. The most important goals for research on these energy storage systems are to improve their cost per kilowatt-hour ($/kWh), specific energy (kWh/kg), cycle life, and calendar life. These storage systems also have to maintain adequate specific power (kW/kg) even at low states of charge.

Irrespective of the propulsion technology, reducing the mass of a vehicle for a given mission will, with no other design changes, have the effect of reducing fuel consumption and increasing acceleration performance. However, to achieve the vehicle goals of the Partnership, any such mass reduction must be accomplished without compromising safety or overall vehicle utility. To accomplish significant weight reductions, several materials, including aluminum, high-strength steel (HSS), and carbon-fiber-reinforced polymer composites have been considered for replacing a large part of the (mostly) mild steel currently used. Other material substitutions, such as cast magnesium, in other vehicle components could further decrease vehicle weight. The challenge for all of these potential material substitutions is to reduce their cost.

The following sections discuss in more detail the research approaches and issues associated with each of these technologies.

ADVANCED COMBUSTION, EMISSIONS CONTROL, AND HYDROCARBON FUELS

Introduction

Even the most optimistic scenario for introducing fuel cell vehicles into the market requires several decades before market penetration becomes sufficient to have a measurable impact on petroleum consumption and CO_2 emissions. During this transition the dominant powerplant for mobility systems will continue to be ICEs fueled with a hydrocarbon (gasoline, diesel fuel, or biofuel). Consequently, it is important to maintain an active R&D program at all levels of industry, academia, and government. The near-term introduction of such technologies could reduce the rate at which petroleum consumption for transportation grows during the transition to alternative powerplants and powertrain configurations.

Adding electric components to a vehicle powertrain opens new ways to improve fuel economy. HEVs are being marketed today, and PHEVs, which offer greater vehicle range from the electricity stored in the onboard battery and can benefit from off-peak battery recharging, are being actively pursued. In hybrid vehicles, the ICE can be operated more efficiently than it is with conventional mechanical or hydromechanical transmissions. Obviously, any improvements to the ICE will carry over directly into incremental improvements of these electrified vehicles.

Maximizing efficiency under the constraint of meeting stringent emission standards complicates engine-powertrain systems. Efficiency improvements require understanding and precise control of every aspect of the engine operation as well as optimization of every component and its interaction within the powertrain system. The technologies currently being pursued with the intent of introducing them to the market for both diesel- and gasoline-fueled vehicles—for example, low-temperature combustion (LTC) and lean exhaust aftertreatment—tax the limits of the engine and powertrain community's fundamental understanding of the controlling thermochemical processes. A greater knowledge of the controlling thermochemical phenomena will accelerate the introduction of these technologies into the marketplace.

The urgent need to get vehicles with lower fuel consumption to market has spurred the effort to incorporate new technologies such as LTC or new lean exhaust aftertreatment systems into vehicle powertrains. After research is completed, system optimization through prototyping and the development of manufacturing procedures can take several years before the new technology reaches production. This time pressure has unified the traditionally separate pursuits of "research" and "development." Researchers are now pursuing enhanced fundamental understanding of the controlling thermochemistry at the same time as engineers are investigating concept systems. Sophisticated research tools such as laser-based optical diagnostics, the detailed identification of chemical kinetic mechanisms and their subsequent simplification, and the implementation of advanced three-dimensional computational fluid dynamics (CFD) are being used to expand the fundamental knowledge base and also being applied to potential configurations for prototype engines and powertrains.

To successfully conduct such a program requires close coordination among industry, government laboratories, and academia. In the opinion of the committee, the advanced combustion and emission control technical team is doing a good job with this close coordination. The organizational structure of their activities involves memoranda of understanding (MOUs) between companies and government labs, working group meetings, regular intergroup reviews, and an annual peer-reviewed research meeting. The committee is pleased with the responses of the technical team to the recommendations made in the Phase 1 report. In particular the team has succeeded in involving the energy companies in its programs.

The energy companies are now actively engaged, and a program known as Fuels for Advanced Combustion Engines has been organized under the Coordinating Research Council. This program has the objective of providing a set of research fuels to discern fuel effects on LTC. The technical team is making good use of its resources. Requests for proposals (RFPs) have been issued for an E85-

optimized engine, for enabling technology for fuels and lubricants, and for clean combustion and fuels co-development.

Funding

FY07 funding for the advanced combustion and emission control technical team was $20.7 million, with the same amount requested for FY08. A breakdown of how the funding was dispensed to different organizations and technologies is shown in Figure 3-1.

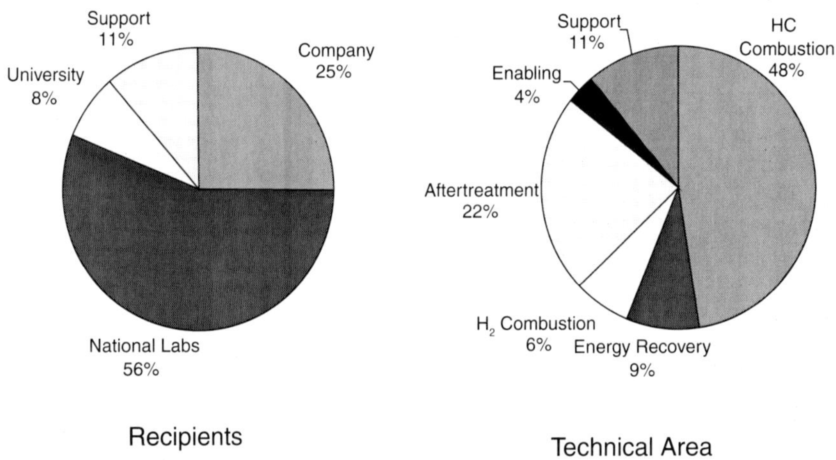

FIGURE 3-1 Distribution of DOE FY06 funding for the advanced combustion and emission control technical team. SOURCE: R. Peterson and K. Howden, DOE, "Advanced combustion and emission control," Presentation to the committee on March 1, 2007.

Goals and Targets

The technical targets and roadmap remain the same as reported previously (NRC, 2005; DOE, 2004). To briefly summarize, the Partnership expects to achieve by 2010 an engine thermal efficiency of 45 percent, with a cost under $30/kW, while meeting Tier 2, Bin 5, emissions. These are very challenging targets. The technical team is working hard to achieve them and is making progress.

The focus of the research continues to be lean-burn, direct-injection engines for both diesel- and gasoline-fueled vehicles. Within this broad area, specific research areas include the following:

- Low-temperature combustion
  —Control,
  —Expanding the load range,
  —Coupling to fuel characteristics,
  —Transient operation, and
  —Combustion mode switching.
- Aftertreatment
  —Modeling of a diesel particulate filter,
  —Lean NO_x traps,
  —Selective catalytic reduction (SCR) NO_x reduction, and
  —Identification of a potential catalyst for hydrocarbon NO_x catalysis.
- Tool development
  —Improved CFD capabilities,
  —Improved diagnostics capabilities, and
  —Comparison of CFD and experimental results.

Within each of the research areas discernable progress has been made. Highlights are documented in "FreedomCAR and Fuel Partnership 2005—Highlights of Technical Accomplishments" and can be found at the USCAR Web site: <www.uscar.org/commands/files_download.php?files_id=95>.

Significant Barriers to Achieving Success

The technologies being pursued by the advanced combustion and emission control technical team are very sophisticated. Making these technologies work in an engine-powertrain system under a range of operating conditions is very challenging. It pushes all the fundamental boundaries of understanding within the combustion and powertrain community. For example, trying to expand the LTC operating condition from an optimal operating point quickly leads to excessive unburned HC and carbon monoxide (CO) emissions, with an attendant reduction in combustion efficiency. The increase in HC and CO poses an additional problem because the exhaust temperatures during LTC are usually below the typical light-off temperatures of current catalysts. Extended-range catalytic converters to facilitate engine operation approaching the edge of the combustion stability regime may solve this problem.

Expanding the LTC operating range of an engine will require optimal matching of the fuel, its distribution within the cylinder, the fluid mechanic mixing, and the temporal and spatial evolution of thermodynamic states within the cylinder. Combustion mode switching between regular spark or diesel combustion and LTC may affect emissions, and the emission behavior of the engine during transient operation within LTC operation is not well understood and could be a serious issue.

Similar statements about the technical complexity of the problems can be made for virtually all the technologies being investigated by the advanced combustion and emission control technical team.

A critical prerequisite for achieving success is expanding the knowledge base of the processes that influence the performance of advanced combustion and emission reduction technologies and their interaction with the fuel being used as the energy carrier.

Conclusions and Recommendations

In light of the FreedomCAR and Fuel Partnership objectives, the funding and work allocation for continued development of the ICE and vehicle electrification seem appropriate. The advanced combustion and emission control technical team is doing a good job of maintaining a close and constructive working relationship with the stakeholders in the auto and energy communities. Since the distinction between research and development has blurred, it is critical for the technical team to maintain this collaboration and make it even stronger. The international competition is fierce, so maintaining a presence within that community and an awareness of technological developments outside the United States continue to be important to establish benchmarks and grow the knowledge base.

The largest barrier to implementing advanced combustion, aftertreatment, and fuel technologies is an insufficient knowledge base. Not only topic-specific understanding but also understanding the system-level interactions between the energy carrier, the energy release process, and the final emission cleanup is critical to continued improvement of the ICE powertrain. Continued close collaboration between DOE and industry is necessary to allow these technologies to transition into industrial laboratories and to identify the areas where enhanced understanding will be most beneficial.

Recommendation. The Partnership should formulate and implement a clear set of criteria to identify and provide support to ICE combustion and emission control projects that are precompetitive and show potential for improvements well beyond those currently being developed by industry.

Recommendation. DOE should actively encourage collaborations among the national laboratories, industry, and academia to more effectively direct research efforts to areas where enhanced fundamental understanding is most needed.

Transient LTC engine operation and combustion mode switching between conventional combustion and LTC could have significant impact on total vehicle drive cycle emissions and the necessary operating domain for catalytic exhaust treatment systems. This could also be relevant to the emissions for PHEVs, where the engine may be off for long periods of time before being started and engaged into the vehicle's powertrain.

Recommendation. The Partnership should investigate the impact on emissions of combustion mode switching and transient operation with LTC.

As the engine is made more efficient and exhaust thermal energy is used for advanced turbocharging and the aftertreatment systems, the final exhaust temperature will become lower. This will reduce the theoretical maximum thermal efficiency of heat engine exhaust recovery systems. Given the projected conversion efficiency of the exhaust heat recovery systems under investigation, the committee questions how much exhaust energy can actually be recovered. In addition, it seems likely that the cost per kilowatt of the heat recovery systems will be high.

Recommendation. The Partnership should perform a detailed analysis of the potential improvement in efficiency and the cost effectiveness of the exhaust gas heat recovery effort and make a go/no-go decision about this work.

FUEL CELLS

Introduction

Hydrogen-based fuel cell powerplants promise to be one of the most efficient and least polluting way to power personal transportation vehicles while providing the potential for meeting the Partnership's major goals. Consequently, the advancement of fuel cell technologies to the point where performance and costs can be compatible with mass-manufactured automobiles is a key element of the FreedomCAR and Fuel Partnership. Over the course of the Partnership, DOE-sponsored fuel cell activities have contributed to solid advances in many of the performance and engineering metrics as well as reductions in projected costs as described in the following section. Current projects are focused on advancing the science and engineering of the high-risk technical challenges that remain, including performance, durability, and lifetime. Initiatives that can lead to cost reductions through materials advancements, new concepts, or simplification of the engineering are also under way and are expected to eventually lead to a mass-manufactured product.

Current Status of Key Parameters

Fuel cell stack life currently limits the overall demonstrated powerplant durability to only about one fourth of what is needed to meet the performance targets set forth by the Partnership. A major reduction in stack life occurs in actual vehicle applications because of the many stops and starts and transients with vehicle operations, fuel composition, and related phenomena when compared to what is observed with the testing methods and conditions in laboratory development work.

In addition, as laboratory fuel cell stack lifetimes lengthen, new failure modes are surfacing and must be better understood and resolved. One such example is platinum catalyst dissolution, which impairs long-term performance. The prompt resolution of these and new failure modes, as they are discovered, is critical to achieving 2010 and 2015 targets.

Projected costs for high-volume (500,000 units/yr) fuel cell powerplant production are currently approximately \$100/kW for relatively proven technologies and about \$67/kW for a newer technology compared to the 2015 target of \$30/kW (see Figure 3-2; also see James and Kalinoski, 2007; Lasher, 2007). The latter estimate includes recent advances by 3M (Debe, 2007) in membrane and electrode technology that result in lower projected costs. These new technologies are quite promising, yet there is still a need to demonstrate that laboratory performance can translate to similar results in full stacks and then, ultimately, in vehicle tests. It should also be noted that the manufacturing and supply chains have not been fully established to date because the technology continues to evolve. The numerous assumptions that underlie the aforementioned cost projections may have to change as the development process proceeds.

The cost of the platinum catalyst used in the membrane electrode assembly (MEA) represents approximately 57 percent of projected stack costs (Lasher, 2007). The platinum metal contained in the electrode alone accounts for most of

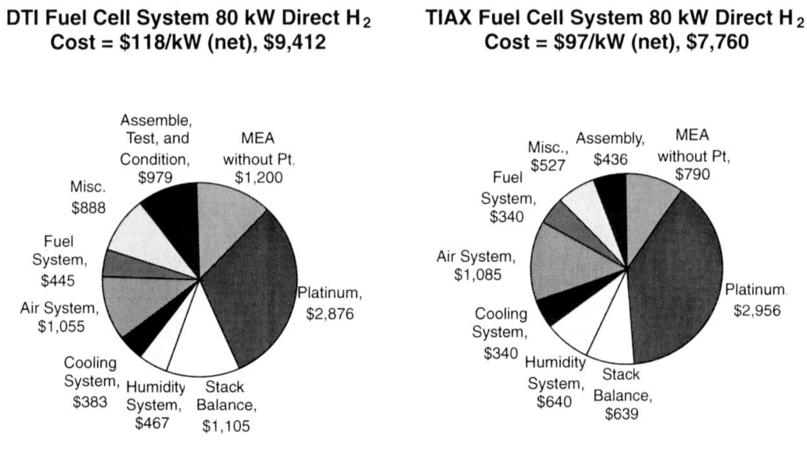

FIGURE 3-2 Two estimates of 2006 costs for fuel cell systems. The differences between the DTI and TIAX estimates are (1) the cost of the MEA and seals in stack balance of plant and (2) DTI included test and conditioning in its estimate. The 2015 cost target is \$30/kW, for a total cost of \$2,400. SOURCE: D. Tran and K. Epping, FreedomCAR fuel cell tech team, Presentation to the committee on March 1, 2007.

the cost of the catalyst, and the price of precious metals has risen substantially in recent years. The spot market price was approximately $1,300 per troy ounce in June 2007, 18 percent higher than the price used for estimating stack costs as reported at the 2007 DOE Merit Review. It must be noted that this price is set by market forces and will not be impacted by the Partnership's research program.

Even though the fuel cell stack is the core of a fuel cell power generation system (and its most discussed element), it must be part of a carefully integrated system to achieve performance goals. The design, performance, and integration requirements of the ancillary components of the system depend heavily on the performance and engineering specifications of the stack, which means that the developmental progress of the entire system is highly dependent on the uncertainties and risks associated with meeting the long-term targets for each of these ancillary components. Of these components, onboard hydrogen storage—even though it does not affect power plant performance directly—is probably the most challenging goal. Major issues associated with this goal are discussed in detail in the next section of this report.

The Partnership has achieved a significant improvement in the power density of the fuel cell stack without storage over the years: 440 W/l (2004), 500 W/l (2005), 580 W/l (2006) vs. the DOE target of 650 W/l in 2010 (Epping and Tran, 2007).[1] Progress in power density, including the storage subsystem, has also been achieved (160 W/l in 2006) and is steadily progressing toward the 2010 target (220 W/l).

The DOE's Office of Basic Energy Sciences (BES) sponsors activities and programs at national laboratories and academic institutions that have used solid state and polymer sciences to study the atomic and molecular structure of catalysts and new polymers. These efforts have improved the knowledge base and could result in new and improved materials and processes for fuel cell components (e.g., membranes and electrocatalysts). The impact of these important efforts will most likely not be realized until after 2010.

Lastly, water dynamics still remains challenging in that, if water distribution is mismanaged, it can reduce stack performance and lifetime. The neutron scattering in situ water imaging technique available at the National Institute of Standards and Technology (NIST) has yielded exceptionally valuable insights into water behavior in an operating fuel cell. With the detection equipment installed in the spring of 2007, it is possible to "see" the water not only in the plates but also under dynamic conditions in the MEA and the gas diffusion layer (Jacobson, 2007). DOE's support for the development of new laboratory techniques such as these is commendable.

[1]K. Epping and D. Tran, FreedomCAR Fuel Cell Tech Team, Presentation to the committee on March 1, 2007.

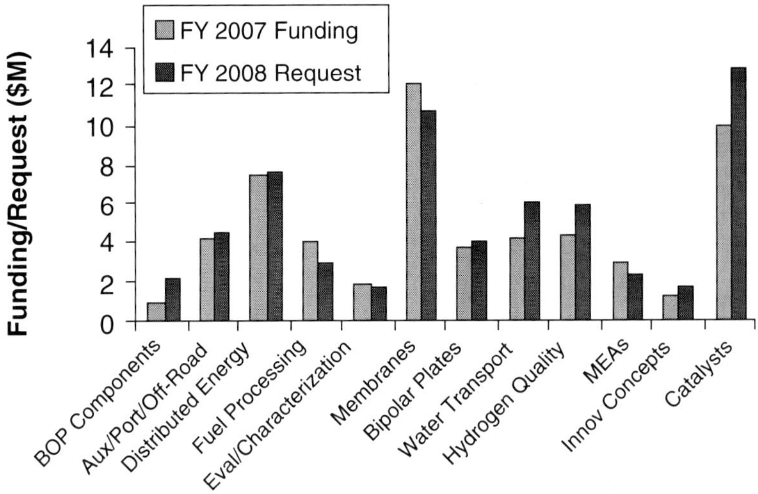

FIGURE 3-3 Fuel cell R&D funding, allocated and requested. SOURCE: Communication between the committee fuel cell subgroup and Kathi Epping and Terry Payne, EERE, on August 28, 2007.

Program Direction and Management

The FreedomCAR and Fuel Partnership's overall requirements (cost, reliability, and performance goals) for the fuel cell system are well established and very challenging.[2] DOE's support is crucial for meeting these requirements. DOE has been actively engaged in funding high-risk R&D for existing programs on a continuing basis and has been able to increase its multiyear development funding for a broad spectrum of technologies and new activities. The budgetary difficulties of FY07 (continuing budget resolution) appear to have abated, and FY08 funding requests have been increased (see Figure 3-3). The potential benefits of this technology and the progress to date justify current spending and increased future spending levels.

The DOE fuel cell program is addressing the high-risk technical elements, and managers have proactively refined some of the near- and longer-term targets that needed refinement. If the 2010 and 2015 goals of the program are to be met, some recently funded R&D initiatives will have to contribute to the timely resolution of many of the remaining technical challenges. Since DOE awarded about $100 million in the fall of 2006 for such development, the research direction and priorities have already been set for the next few years. To gain further valuable

[2]See <http://www1.eere.energy.gov/hydrogenandfuelcells/mypp/pdfs/fuel_cells.pdf>.

insight into the resolution of engineering issues, DOE funded and supported the Learning Demonstration Program using on-road, fuel-cell-powered vehicles supplied by the automotive partners. This program will help to further identify and quantify state-of-the-art capabilities and deficiencies of fuel cell power generation systems (see Chapter 2).

Assessment of Progress and Key Achievements

Steady progress in fundamental research on polymers and catalysts has been demonstrated. However, it is difficult to assess this progress in terms of the program targets until such time as the technologies are demonstrated on-board a vehicle or in a laboratory situation where vehicle operation can be accurately simulated. The 2010 goal for power system density includes the fuel cell power plant as well as the complete hydrogen storage and delivery subsystems. For this reason and because of the apparent difficulty in meeting storage density targets, it seems unlikely that power system goals can be met, even with substantial progress on the fuel cell power plant.

Although the cost estimates for high volumes (500,000 units/yr) show steady progress, the accuracy of these estimates is limited, because many of the final technologies and manufacturing processes are still evolving or are unproven. Even with such uncertainties, however, the estimates allow benchmarking and assessing progress. The accuracy of these cost estimates will improve as advanced technologies move out of the laboratory, complete systems are better defined, and performance is demonstrated.

BES materials R&D has started to yield fundamental knowledge about the membrane and electrodes. Such work should continue in force. The delay in initiating funding of selected aspects of the BES program (see, for example, Kung, 2007) may have a negative impact on the program in the longer term.

Significant Barriers and Key Issues

On the technical front, a number of barriers remain that were addressed in the Phase 1 review. Membrane and catalyst lifetimes, reliability, and durability remain problematic, as do barriers within other subsystems. Water management is still a challenging operating parameter as it impacts membrane conductivity, freeze protection, and electrode performance, and ultimately balance of plant complexity. Too little or too much water can cause drying out or flooding. Platinum loading needs further reduction, especially as platinum spot market costs remain high and unpredictable. In addition, new issues have started to emerge and will have to be addressed, including catalyst loss due to platinum dissolution into the membrane. Others include the possible impact on performance and system costs associated with intake air quality and coolants; at the same time, electrode composition is being reevaluated with respect to durability.

The technical approaches should be reviewed in light of recent knowledge and advancements gained from earlier funded efforts and vehicle testing. Sensitivity studies should be conducted to determine the consequences of variability in operating parameters on performance, cost, and system design, as well as the programmatic issues of not meeting various targets.

To meet the 2015 goals, the currently funded and newly awarded programs must come up with viable technical solutions in the next 3 to 5 years. Funding must remain intact, and delays in funding certain critical efforts must be minimized. Allocated budgets should be reassessed and reallocated as necessary, with the emphasis on the highest priorities. For example, the possibility of making up for the delays in funding new membrane research (Kung, 2007) at the expense of less critical programs should be considered. There can be value in funding stationary programs, which could help establish the supply chain vendor base and contribute to solving key technical issues. Nevertheless, stationary opportunities and markets are still emerging, and stationary and vehicle technologies are often significantly different, so overlap may be minimal. Funding for stationary programs under the Partnership should be reassessed to ensure that it is used only for technologies that have a clear value to vehicular applications.

With membrane and electrode technology still under development and stack designs still being enhanced, it is difficult for the supply base or systems developers to build manufacturing operations and invest in fixed assets at this time. It is commendable that the DOE has identified component and stack manufacturing—in particular, the development of manufacturing models and processes—as areas that will need to be addressed later on. With the design uncertainties that exist, however, any funding for manufacturing initiatives must be restricted to generalities at this time rather than applied to specific component designs and materials.

The purity of the hydrogen fuel entering the stack remains a significant factor for both fuel cell performance and durability and is currently being addressed. The same should be done for impurities in the air entering the stack.

Recommendations

Recommendation. The Partnership should conduct sensitivity analyses on key fuel cell targets to determine the trade-offs and tolerances in engineering specifications allowable while still meeting fuel cell vehicle engineering requirements.

Recommendation. The Partnership should reassess the current allocation of funding within the fuel cell program and reallocate it as appropriate, in order to prioritize and emphasize R&D that addresses the most critical barriers. In particular, the Partnership should give membranes, catalysts, electrodes, and modes of operation the highest priority. In particular, it should also

- Place greater emphasis on science and engineering at the cell level and, from a systems perspective, on integration and subcomponent interactions;
- Reduce research on carbon-based supported catalysts in favor of developing carbon-free electrocatalysts;
- Ensure that BES funding of membranes, catalysts, and electrodes remains a high priority of the program; and
- Apply go/no-go decision making to stationary fuel cell system initiatives that are not directly related to transportation technologies.

ONBOARD HYDROGEN STORAGE

Background

Storing enough hydrogen on board the vehicle to provide a 300-mile driving range while simultaneously meeting weight, volume, and cost targets continues to be very challenging. At this time, the only demonstrated complete, workable storage systems use highly compressed hydrogen gas or liquid, but they are unlikely to meet the 2010, much less the 2015, targets. The committee therefore believes that the research activities of the FreedomCAR and Fuel Partnership are appropriate in view of the need for new materials, reduced cost, and thermally integrated processes for efficiently storing and releasing clean hydrogen onboard the vehicle.

Until a no-go decision was made in 2003, DOE focused on R&D for onboard fuel processors to produce hydrogen from stored-onboard liquid feedstocks, especially methanol or gasoline. As an alternative, DOE recently initiated a broad-based R&D program on the storage of hydrogen, which can be utilized onboard the vehicle. Hydrogen storage had been explored in the past, but mostly for niche applications. This DOE-sponsored program is the first-ever effort with so many researchers pursuing alternative technologies simultaneously with cost and performance targets. As this effort began only recently, hydrogen storage technologies lag other technologies for hydrogen fuel cell vehicles. This delay could impact the overall schedule for the program.

The targets and the time line for technology development included in the Hydrogen Storage Technologies Roadmap are aggressive, particularly when one considers that all targets must be met simultaneously. These include targets for volumetric and gravimetric storage, cost, thermal management (hydrogen flow rate requirements and energy efficiency targets), and a minimum number of fill/discharge cycles. The initial strategy is to explore a diverse portfolio of candidate materials that could potentially lead to acceptable complete storage systems. Thus, appropriate materials are necessary but not sufficient to resolve the storage issue.

The hydrogen storage program is currently organized around three hydrogen storage centers of excellence (COEs) and a number of independent projects. The COEs and independent projects include work by 40 universities, 15 companies, and 10 federal laboratories. The budget for hydrogen storage activities was $26 million for FY06 and $34.6 million for FY07. The requested budget for FY08 is $43.9 million. Lack of earmarks in the FY07 Continuing Resolution has benefited the program. BES funding for hydrogen storage was approximately $11 million and $12 million for FY06 and FY07, respectively; the request for FY08 was about $17 million.

Current Status of Hydrogen Storage with Respect to the Targets

The OEMs—namely, the automotive companies—are currently and presumably for the foreseeable future using compressed gas storage in their demonstration vehicles, even though the volumetric storage densities (as well as costs) will not meet the 2010 or 2015 targets. Liquid hydrogen storage and 700-bar (10,000-psi) high-pressure storage have the highest volumetric and gravimetric storage densities demonstrated, respectively, but fall significantly short of the 2010 and 2015 targets for both performance and cost. In addition, liquid hydrogen is stored at –253°C (–423°F), which introduces many additional problems. The compressed gas tanks being utilized by the automobile OEMs are made from carbon fibers wound around either metal liner tanks (type 3) or high-density-polyethylene liner tanks (type 4), which are bonded with resin. The carbon fibers make up more than half of the system weight and costs. Further, the maximum allowable temperature (85°C) limits refill rates for storage at 700 bar (10,000 psi). Therefore, one DOE activity is an effort to reduce carbon fiber costs.

From 2004 to 2006, the hydrogen storage COEs made significant progress in identifying materials with increased hydrogen storage capacity. Yet the materials identified to date are still far below the target levels for net (usable hydrogen, accounting for all energy losses during filling and release) storage of hydrogen onboard the vehicle. The DOE targets for onboard hydrogen storage are 6 and 9 weight percent for the entire system (Satyapal, 2007) for 2010 and 2015, respectively. This goal is often confused with the chemical weight percent of hydrogen in the material. Given that the balance of plant for the system will be at least 50 percent of the system, the material weight percent of hydrogen should be halved in order to judge it against the 2010 and 2015 system goals. Of the 24 sample materials presented at the Hydrogen Program Merit Review (Satyapal, 2007), 16 are equal to or more than 6 weight percent, while 8 materials fall below this value for the material alone.

Accounting for the balance of plant for the system, only one material will meet the 2010 goal, but this material will not meet the 2015 goal. The target for 2015 was set to provide a driving range equivalent to 300 miles between fill-ups and make the hydrogen storage system approximately the same size and weight as a gasoline system. Even if materials are identified that meet the weight requirement, they may be incompatible with other targets due to energy or temperature requirements and/or absorption/release rates that are too low. Furthermore, few if any have progressed to the point of complete system design and evaluation. Driving ranges of 300 miles can be achieved with weight fractions of less than 9 percent given innovative overall vehicle designs that decrease the hydrogen storage capacity requirements or increase the space available for fuel storage. Many of the initial round of projects (normally 3- to 5-year contracts) will be up for renewal or discontinuation within the next year. Projects focused on materials not having the potential for meeting targets need strong justification for renewal.

Assessment of Progress and Key Achievements

The three broad classes of materials for hydrogen storage being screened are sorbents, chemical hydrides, and metal hydrides. During the past 2 years several very promising approaches to hydrogen storage have been pursued, including (1) hydrogen storage in engineered metal organic frameworks with very high surface areas that adsorb significant amounts of hydrogen at cryogenic temperatures, (2) progress in lowering the temperature of hydrogen release from ammonia borane with the addition of lithium amide, (3) exploration of hydrogen spillover as a route to ambient temperature storage, and (4) the development of computational techniques/methodology for predicting the thermodynamics of complex, multicomponent hydride systems. Additionally, organic liquid systems are being designed that minimize energy losses.

DOE should be commended for sponsoring a demonstration of a hydrogen storage system with a prototype based on $NaAlH_4$. A significant finding was that the balance of system weight was about 50 percent of the total system weight. This finding provides information on what may be required in a practical system for heat exchangers, vessels, manifolds, and other components.

Significant Barriers and Issues That Need to Be Addressed

It is not clear at this time that a suitable hydrogen storage material will be identified that can meet program goals and timing targets. Without a suitable hydrogen storage material, widespread deployment of a hydrogen fuel cell vehicle for transportation would be a market risk. The hydrogen storage targets for 2010 and 2015, set at the start of the FreedomCAR and Fuel Partnership, were chosen based on attaining a 300-mile vehicle driving range and making assumptions about overall powerplant and vehicle characteristics. Storage of liquid hydrogen or hydrogen at high pressure can apparently meet the 300-mile range target but not other targets. Alternative storage materials identified to date have not been shown to meet any of the storage system capacity targets.

The committee believes that the stringent storage targets should be kept for approximately 4 more years (until 2011). At that time, (1) the progress toward the 2010 system goals will be known (and they will probably not be met), and the realism of the 2015 goals can be reassessed; (2) perhaps all of the major goals (not just storage) should be reexamined to see how the vehicle optimizes; and (3) the Partnership will be able to make trade-offs between storage, structural materials, batteries, fuel cells, and other major subsystems of the vehicle. Delaying the assessment of targets to 2011 will also allow a second round of 3-year contracts to be completed.

Thermal management is also important. For all the materials considered, energy is required to either absorb or release the hydrogen, with rates for these processes depending on temperature and pressure. Ideally, the waste heat from the fuel cell (currently around a temperature of 85°C) would be used in the fueling process, since otherwise additional fuel would be consumed, lowering overall efficiency. Accordingly, systems must be thermally compatible as well as carefully integrated to allow both hydrogen rates and efficiency targets to be realized.

A proposed new hydrogen storage COE on systems analysis was noted in the presentations to the committee. This COE should help with downselection (go/no-go). It should also model and evaluate conceptual systems to ensure that system targets are met and materials and thermal balance and dynamics are compatible with the fuel cell stack and subsystems.

The hydrogen storage program organized in BES is also of critical importance given the need for the discovering new materials and for understanding material properties that might lead to the improvement of current promising candidate materials. New starts in the BES program were severely hampered by funding limitations under the 2007 Continuing Resolution. The Partnership should endorse a robust program in the basic sciences. The committee is disappointed that BES's request for increased FY07 basic hydrogen research funding was not fully authorized under the joint congressional for FY07.

Response to Recommendations on Hydrogen Storage from the Phase 1 Report

Communicating the status of suitable hydrogen storage materials and systems to those who set policy is essential in view of the criticality of hydrogen storage to the long-term goals of the Partnership. In the Phase 1 report, Recommendation 3-9 recommended reporting the status of the technology annually to all program participants, DOE, and the Congress. The following were published during the last 2 years of the program: (1) reports to all the technical teams, (2) a report to the Fuel Operating Group (high-level company managers), (3) report to the Executive Strategy Group (automotive company vice presidents and DOE Undersecretary David Garman), and (4) reports to Congress on the state of the research program in hydrogen storage (in the form of testimony). There was also participation in the International Partnership for the Hydrogen Economy. The committee is satisfied with these reporting and other activities.

In view of the size of the hydrogen storage program and the diversity of investigative approaches, participation in many technical conferences and in the yearly contractors' program review meeting are critical means for communicating progress and evaluating the work across the program as a whole. These report opportunities focus dialogue on the most promising approaches and foster the generation of new ideas and discoveries.

Recommendation 3-7 recommended checking progress in terms of go/no-go decisions. Accordingly, DOE established a go/no-go process for determining which materials are worthy of continued R&D. The process entails criteria development, data gathering, and review of results, followed by evaluation and final decisions by DOE. It ensures that time and resources are devoted to materials that exhibit the most promise, and it enables new materials to be cycled into the program for study and evaluation. This evaluation procedure is judged by the committee to be adequate. Certainly the screening out of unworthy candidate materials should be an ongoing process requiring little formality.

A go/no-go decision was conducted on pure single-wall carbon nanotubes. The hydrogen storage capacity of pure carbon nanotubes was judged to be far short of the storage targets, so work was, appropriately, discontinued following a no-go evaluation. Cryocompressed storage was assessed, and a go/no-go decision was planned for fall 2007. A go/no-go decision on the $NaBH_4$ is planned for the same time frame; in addition, a downselection is expected on a reversible metal hydride.

When the Phase 1 report was issued, the hydrogen storage component of the FreedomCAR and Fuel Partnership was just being established. The COE approach for project organization and project management had not been tested. Now, 2 years into the research, three COEs for hydrogen storage ($5 million to $6 million per year each) have been established. The current centers, which are planned for 5 years (to FY10) are as follows: the metal hydride center, the hydrogen sorption center, and the chemical hydrogen center. The Phase 1 review committee asked in Recommendation 3-8 that the existing COEs be evaluated before extending the approach to other areas. During the past year the program benefits and difficulties have been identified in a self-evaluation process that elicits responses from center members. Benefits identified include a critical mass of researchers with similar interests, collaborations and sharing of equipment, common methods for evaluation of new materials, with testing done at the Southwest Research Institute, rapid communication, and best safety practices. Participants are aware of hydrogen storage targets and criteria. The oversight of the program provided by DOE sets a standard for communicating progress through contractor meetings. One downside was limited flexibility of research. Also, university partners found it somewhat difficult to adapt to milestone-driven work. Based on this analysis, the committee finds that the COE system is working well.

Appropriate Federal Role

Sponsorship of the hydrogen storage component of the FreedomCAR and Fuel Partnership is an appropriate role for the federal government. Work on hydrogen storage has been advanced significantly through the large increase in the number of qualified researchers, the sharp focus on common goals, enhanced communication among participants, and accountability for results.

Recommendations

The hydrogen storage program has reported significant progress during the past 2 years, yet results reported to date are still far short of the 2010 and 2015 system targets.

Recommendation. The hydrogen storage program should continue to be supported by the Partnership at a high level since finding a suitable storage material is critical to fulfillment of the vision for the hydrogen economy. Both basic and applied research should be conducted.

At the beginning of the hydrogen storage program a wide net was cast in search of suitable hydrogen storage materials. It is now becoming clear that many approaches and materials may not be worth pursuing, even at a basic level.

Recommendation. The Partnership should rebalance the R&D program for hydrogen storage to shift resources to the more promising approaches as knowledge is gained. The new systems engineering center of excellence should look at all of the system requirements simultaneously, not just the system weight percent storage goal, and guide this rebalancing.

Recommendation. In the event that no onboard hydrogen systems are found that are projected to meet targets, the Partnership should perform appropriate studies to determine the risks and consequences of relying on pressurized hydrogen storage. They should include production and delivery issues as well as effects on vehicle performance, safety, and costs.

Recommendation. The Partnership should pursue research leading to lower costs for high-quality carbon fibers and bonding materials that would allow higher operating temperatures for compressed hydrogen gas storage.

Recommendation. The Partnership should maintain a strong basic research activity on hydrogen storage. New hydrogen storage concepts should continue to be supported by the Office of Basic Energy Sciences.

ELECTROCHEMICAL ENERGY STORAGE

Introduction

Electrochemical energy storage technologies are critical to the development of HEVs, which would play at least a key transitional role in achieving the FreedomCar and Fuel Partnership's long-term goal of clean and sustainable energy for transportation systems and may become central to achieving these goals if development of fuel cells and fuel cell vehicles is not sufficiently successful to result in their large-scale commercial introduction. For a long time, the FreedomCAR and Vehicle Technologies (FCVT) program has supported the development of advanced batteries and ultracapacitors for lightweight and heavy-duty vehicles, with particular focus on advancing the development and commercialization of HEVs, hybrid fuel cell vehicles (HFCVs), and battery electric vehicles (EVs). In response to the President's Advanced Energy Initiative, which he announced in his 2006 State of the Union address, the FCVT began the development of components for plug-in hybrid vehicles (PHEVs), including advanced batteries for this application.

Currently HEVs are a small but growing part of the U.S. automotive market. In 2006 HEV sales accounted for about 1.5 percent of new vehicle sales, and hybrid technology has continued to penetrate across a variety of vehicle platforms. In 2006, 10 different models of HEVs were available, and another 8 models are expected to be introduced by 2009. In addition, a hybrid version is being provided as an option in several existing models. All the HEVs available at present use a nickel metal hydride battery, and DOE has been involved in the advancement of this technology since the 1990s. However, the nickel metal hydride battery will not meet the long-term FreedomCAR and Fuel Partnership electrochemical energy storage goals for HEVs of 15-year life with 25 kW pulse power and $20/kW by 2010. Thus the FCVT is primarily focused on the development of Li ion batteries for HEV, HFCV, and EV applications.

FCVT has expanded the electrochemical energy storage activity to include PHEVs, with a goal of developing vehicles that can travel about 40 miles on electric energy stored in the battery, which represents about 70 percent of the daily commuting mileage in the United States. PHEVs operate in both electrical and mechanical (as in HEVs) and electric only (as in EVs) modes, and the battery can be recharged from a standard electric outlet. The FCVT efforts are directed at developing PHEV components and systems that could be commercialized sometime between 2016 and 2020. At present, analytical and benchmarking activities are being conducted to determine the benefits and requirements for PHEVs. In February 2007 FCVT released an external draft of the PHEV R&D plan, which was modified and rereleased in June 2007. In April 2007 a request for proposal information (RFPI) was announced by the United States Advanced Battery Consortium (USABC) for the development of advanced high-performance batteries for PHEV application; the RFPI was expected to be awarded later in 2007.

In addition, BES plans to increase its basic research on energy storage technologies. BES held the workshop "Basic Research Needs for Electrical Energy Storage" in March 2007, published an R&D plan in July 2007, and is planning to fund projects in 2008. It will focus on long-term needs, such as basic understanding of materials, interfacial charge transfer, and tools and processes to design new materials. Although the BES mandate on energy storage is broader and longer term, it recognizes the importance of working closely with FCVT on energy storage needs for automotive applications.

FCVT, in collaboration with USABC, manages the technology for electrochemical energy storage. The technology is being developed by battery manufacturers, DOE national laboratories, and universities and through awards under the Small Business Innovation Research (SBIR) program. The effort comprises three subactivities: (1) battery technology development is involved in battery system module development, technology assessment, and benchmark testing; (2) applied battery research focuses on understanding failure and the life-limiting parameters of the Li ion system, which is currently closest to meeting the technical goals; and (3) long-term battery research addresses fundamental understanding of specific electrochemical systems for Li ion batteries. Over the last few years, just under 20 percent of the FCVT budget has been directed at electrochemical energy storage technologies. In FY06, of the $24.4 million total budget, $17.4 million were directed at battery development, $1.4 million at applied battery research, and $4.5 million at long-term research. The total funding for FY07 is $40.8 million, and the FY08 request is $41.8 million; this significant increase over previous years is for the development of PHEV batteries.

Program Status and Assessment

All the HEVs on the market use a nickel metal hydride battery; however, because this electrochemical system has an inherently low specific energy density and uses expensive materials, it will not meet the performance or cost targets of the Partnership. Thus FCVT's focus on the development of a Li ion battery is correct since it has the best potential to meet the long-term goals of the Partnership. The Li ion battery has a higher voltage (>3 V vs. 1.3 V for nickel metal hydride), which is an advantage in building higher-voltage (up to 400 V) automotive power systems, and a higher specific energy density (demonstrated 120 Wh/kg vs. 75 Wh/kg for nickel metal hydride), and the technology is capable of further growth. Tests of the entire battery system show that the Li ion battery will exceed the FreedomCAR and Fuel Partnership 2010 battery system weight and volume goals at the minimum pulse power rating of 25 kW. It is expected that with further improvements the Li ion battery will also meet the weight and volume goals at the maximum pulse power rating of 40 kW.

Significant improvements have also been demonstrated over the last 2 years in other performance parameters of the Li ion battery. The battery meets the cycle life requirement of at least 300,000, and progress has been made in meeting the calendar life target of 15 years (more than 10 years has been demonstrated). Furthermore, the battery will operate over a wider temperature range, and its cold cranking power has been improved.

FCVT and the larger battery community have recognized that safety issues are an important part of battery development, and safety-related issues are being worked on by all subgroups of the Li ion battery development program. This involves understanding the thermochemical and electrochemical stability of the individual materials and single cells and the abuse testing of battery modules. (Extensive testing of the battery component and systems are conducted not only within the window of operation of the battery, such as the voltage, current, temperature, and so forth, but also outside this window, and such testing is called "abuse" testing.) Over the last few years the abuse tolerance of the battery has continued to improve. Many abuse-related issues can also be addressed by external electronic control; however, it is imperative that researchers continue to look for battery chemistries that are resistant to voltage or thermal abuse.

The recent rash of fires in laptops using Li ion batteries may have left the public with the idea that these batteries cannot be made safe. This perception should be balanced against the reality that such failures are very rare and Li ion batteries are still in the early stage of development. In the past, generally during early development, today's safe lead acid and nickel metal hydride batteries were perceived to have safety issues. A safety concern has also been raised about the scale-up of Li ion batteries from the AA type cells used in cell phones and laptops to the larger cells used in automotive applications. Large Li ion batteries and other lithium-metal-based batteries have been safely deployed in military and space applications and, in Japan, in some trucks. The FCVT should continue to be forthright and transparent about all safety-related concerns, tests, and results, not only by making technical improvements but also by correcting any misperceptions.

Performance improvements and the abuse tolerance issues of Li ion batteries are being addressed at all levels of development. Similar systems (nickel/cobalt oxide–carbon or manganese oxide–carbon) are being investigated at the level of the entire battery system by the battery technology development subactivity; single-cell performance is being investigated by the national laboratories as a part of the applied battery research subactivity; and a basic understanding of components and materials is being sought at universities in the long-term research subactivity. Similarly, more stable anodes, such as lithium iron phosphate ($LiFePO_4$), and alternative cathodes, such as nano lithium titanium oxide ($LiTi_{12}O_5$), which prevents the deposition of metallic lithium, are being investigated to improve the abuse tolerance of the Li ion battery system. Again, this investigation is being conducted at the materials level in the long-term battery research subactivity and as cells and batteries in the advanced battery research subactivity. There appears to be coordination of efforts in investigating similar and related materials and systems issues across all three subactivities of the Li ion battery development effort, and it is hoped that there is close communications between groups to accelerate progress.

Although significant progress has been demonstrated in the performance of the Li ion battery, the cost of this battery remains a major barrier to its introduction in HEV applications. Currently, the cost for the HEV battery in volumes of 100,000 units per year is estimated to be between \$750 and \$900, which is almost twice the FreedomCAR's 2010 target of \$500. However, in comparison to the 2004 cost estimate of \$1,200, there has been significant reduction in cost. Cost is a critical factor in the introduction of Li ion batteries in HEV applications since they will replace existing and presumably lower cost nickel metal hydride batteries. Going forward, it is generally expected that the cost of the mature nickel metal hydride battery will be tied to the commodity price of nickel, while the evolving Li ion battery technology, which can use a variety of lower cost materials, will eventually become much cheaper. In fact the main improvement in the cost of the Li ion battery over the last 2 years comes from replacing the expensive $LiCoO_2$ by cheaper $LiMn_2O_4$ or $LiFePO_4$ for the cathode in the battery. Another expensive material in Li ion batteries is the microporous separator, and FCVT has funded two programs to reduce the cost of the separator by half, to about \$1/m^2. FCVT should be commended for recognizing that cost reduction will primarily be achieved by investigating alternative low cost materials and aggressively pursuing various combinations of materials for the Li ion battery. Furthermore, the performance and abuse tolerance of these potentially lower cost materials are being simultaneously studied by the various subactivities at all levels, from basic research on understanding the materials themselves to research at the level of cells to determine their life-limiting processes.

The fact that Li ion batteries can be made from a variety of materials is at one and the same time both the strength of this technology and its difficulty—namely, developing a viable commercial product. On the one hand, the variety of materials that can be used to make a battery suggest that there is significant room for increasing the battery's energy density, improving other performance characteristics, and reducing cost. On the other hand, the development of the battery is made more difficult by the wide choice of materials, since not only do the individual materials have to be characterized but each electrochemical couple has to be characterized for its performance, abuse tolerance, and cost at the cell and battery module levels.

Replacing expensive cobalt ($LiCoO_2$) with low-cost manganese ($LiMn_2O_4$) or iron ($LiFePO_4$) could significantly reduce the cost of the battery, but it would still be much higher than the target cost of \$500 per battery. These estimates have a wide window (\$750 to \$900) at a production rate of 100,000 units per year. Thus, larger production volumes may be required to reduce the price of the battery, and a detailed study of the effect of production rate on battery cost should be undertaken. In fact, the cost study should be carried out to at least 500,000 units per year to allow a meaningful comparison against fuel cell production costs. It may also be necessary to revisit the cost target established for the battery and substitute a more realistic target. It is possible that the initial assumptions, both market and technical, may have to be refined to reflect the present market conditions such as gasoline prices. In addition, the greater base of knowledge about materials and processing gained over the last few years should be factored in to obtain a more realistic cost target for these batteries. In any case, the impact of higher battery cost on the cost of an HEV should be determined.

The cost of the battery will play a large role in the eventual success (or otherwise) of the PHEV, which operates in both the HEV mode, requiring high pulse power, and the electric mode, like an EV, where the increase in electric energy required is proportional to the electric mileage requirements. Thus, while an HEV requires a battery delivering only 1.5 to 2 kWh, a PHEV with a 10-mile electric range will require 5-7 kWh of energy from the battery and a PHEV with a 40-mile electric range will require 10-15 kWh (the energy and power requirements can depend on the charge depleting and sustaining modes chosen for the PHEV application). The FCVT's cost goal for the high-power HEV battery is $250/kWh; its goals for high-energy EV batteries are $150/kWh in the short term and $100/kWh in the long term. Since the PHEV battery needs both high power and high energy and the ratio of the power to energy changes with the desired electric range, a new normalized cost requirement needs to be established for PHEVs. Although the PHEV cost goal has not been finalized, the draft short-term goal is a 10-mile electric range at a battery cost of $300/kWh and the draft long-term goal (2016) is a 40-mile (or more) electric range at a battery cost of $200/kWh. The energy and power required from the battery for the PHEV application depend on the electric range and the relative charge-depleting and charge-sustaining modes. Some performance and cost goals that may be under consideration are listed in Table 3-1.

The recognition of the potential benefit of PHEVs, a reduction in petroleum consumption, has led to growing support from the government and, in 2007, to $27.5 million in funding for the development of PHEV-related components and systems. However, progress has been extremely slow, and although the first discussions on a PHEV program began in May 2006, FCVT and its partners were unable to finalize the PHEV R&D plan by fall 2007. The latest draft plan states that the PHEV program goal and the development targets are expected to be completed sometime in 2008. There is a serious lack of urgency in executing this important plan, and the reasons for the delay are not clear. Furthermore, DOE is delivering inconsistent messages. On the one hand, the PHEV program has been presented as one of the elements in the transition to hydrogen-driven vehicles that are to be ready for a commercialization decision in 2015. On the other hand, the June 2007 draft plan for PHEVs calls for commercialization between 2016 and 2020. These mixed messages can only cause confusion among the interested parties. It is very important that DOE present a single and consistent R&D plan for PHEVs immediately.

TABLE 3-1 USABC Goals for Advanced Batteries for PHEVs

Characteristics at End of Life	Unit	High Power/Energy Ratio Battery	High Energy/ Power Ratio Battery
Reference equivalent electric range	miles	10	40
Peak pulse discharge power at 2 sec/10 sec	kW	50/45	46/38
Peak Regen pulse power (10 sec)	kW	30	25
Available energy for CD mode, 10 kW rate	kWh	3.4	11.6
Available energy for CS mode	kWh	0.5	0.3
Minimum round-trip energy efficiency (USABC HEV cycle)	%	90	90
Cold cranking power at −30°C, 2 sec-3 pulses	kW	7	7
CD life/discharge throughput	cycles/MWh	5,000/17	5,000/58
CS HEV cycle life, 50 Wh profile	cycle	300,000	300,000
Calendar life, 35°C	year	15	15
Maximum system weight	kg	60	120
Maximum system volume	liter	40	80
Maximum operating voltage	V_{dc}	$>.55 * V_{max}$	$>.55 * V_{max}$
Minimum operating voltage	Wh/day	50	50
System recharge rate at 30°C	kW	1.4 (120 V/15 A)	1.4 (120 V/15 A)
Unassisted operating and charging temperature range	°C	−30 to +52	−30 to +52
Survival temperature range	°C	−46 to +66	−46 to +66
Maximum systems production price, 100,000 units/yr	$	1,700	3,400

NOTE: CD, charge depleting; CS, charge sustaining.
SOURCE: USCAR Request for Proposal Information for Advanced High Performance for Plug-in Electric Vehicle Applications. Available on the Web at <http://www.uscar.org/guest/publications.php>.

Recommendations

Recommendation. The Partnership should conduct a thorough analysis of the cost of the Li ion battery for each application; hybrid electric vehicles (HEVs), PHEVs, battery electric vehicles (EVs), and hydrogen-fueled fuel cell HEVs. The analysis should re-examine the initial assumptions, including those for both market forces and technical issues, and refine them based on recent materials and process costs. It should also determine the effect of increasing production rates for the different systems under development.

Recommendation. The Partnership should significantly intensify its efforts to develop high-energy batteries, particularly newer, higher specific energy electrochemical systems within the long-term battery research subactivity and in close coordination with BES. High-energy batteries provide the surest way to successful batteries for PHEVs.

Recommendation. The Partnership should move forward aggressively with completing and executing its R&D plan for plug-in hybrid electric vehicles.

ELECTRIC PROPULSION, ELECTRICAL SYSTEMS, AND POWER ELECTRONICS

Introduction

The scope of the FreedomCAR and Fuel Partnership includes R&D aimed at commercial advancement of HEVs, fuel cell HEVs, EVs, and PHEVs. Electrical systems in all these types of vehicles consist of electric propulsion systems and power electronics systems, along with appropriate electronic controllers. Electric propulsion systems convert electrical energy from the fuel cell and/or electrochemical energy storage device (e.g., a battery) into propulsive force interfaced to the wheels through appropriate drive trains and vice versa. Power electronic systems are used to convert the electrical energy among various forms (current, voltage, dc, ac, frequency) for energy flow between a fuel cell, and electrochemical energy storage device, a rotating electric machine, an internal combustion engine, and the electric utility. Electric machines and power electronic systems thus form an enabling technology critical for achieving the Partnership's goal of clean and sustainable energy for transportation systems, cutting across the various approaches in both the near term and the long term. The relatively recent emphasis on PHEVs has raised additional technological concerns about interactions between the electric grid and the vehicle. Concomitantly, the budget appropriation for these activities has been essentially steady in recent years (for FY07, \$15.6 million; for FY06, \$13.6 million).

HEVs are gradually becoming established and noticeable in the marketplace, spurred by increasing gasoline prices. Although the share of new vehicles sold accounted for by HEVs is small, their absolute numbers are growing rapidly, and they are the segment with the fastest growth. The electric propulsion and power electronics technologies in these vehicles continue to improve, with new models introduced every year. The technologies of permanent magnet electric motors and power electronic converters are showing aggressive design and rapid progress in their performance. In recognition of the critical nature of these enabling technologies, there has been an increase in activities focused on them.

A majority of the technical activities are being conducted by DOE national laboratories, by universities, and by automotive equipment vendors. During 2006, a comprehensive solicitation aimed at developing advanced technologies was issued and research awards were announced in May 2007. The projects include a significant amount of cost share from industry focused in four areas: (1) high-temperature, three-phase inverters, (2) high-speed motors, (3) integrated traction drive systems, and (4) bidirectional dc-to-dc converters. The goal in all four areas is to reduce the cost, weight, and package size of electric drive and power conversion devices while increasing vehicle efficiency. These projects represent are spending about $33.7 million, including the 50 percent cost sharing by industry.

Program Status

The current generation of HEV devices uses a permanent magnet ac motor, along with insulated gate bipolar transistor (IGBT)-based inverters and dc-dc power converters to manage the power flow among various energy sources and loads, with an appropriate cooling system. Improvements in the cost, weight, and volume of these systems are projected to come from developments in (1) the high-temperature operation of materials, components, and subsystems; (2) high-speed electric machines; (3) novel power converters that minimize the use of capacitors and magnetic elements; and (4) the integration of subsystems and components.

Higher Temperature Operation

Current-generation power conversion devices utilize a secondary coolant loop that operates at a lower temperature than the primary engine coolant loop. Advances in the technology to operate the devices at higher temperatures would lead to improvements in cost, reliability and power density. A variety of projects aimed at this include phase change cooling, compressed air cooling, spray cooling, high-temperature capacitors, high-temperature insulation, high-temperature packaging, wide band-gap materials such as silicon carbide power devices, high-temperature gate drives, high-temperature magnets, and thermal interface materials. While these projects are broadly aligned with the goals of the program, it is unclear whether there is a clear, quantifiable impact on meeting the performance and cost goals of the program.

High-Speed Electric Machines

The current generation of permanent magnet electric machines for vehicle propulsion operate at about 15,000 rpm. For a given power level, the general scaling laws of rotating electric machines lead to a proportionate reduction in the weight of the active materials, copper and iron, as the operating speed is increased. However, the accompanying design trade-offs involved in increasing the speed are related to various electrical and mechanical parameters such as voltage, current, losses, shear stress, fault tolerance, manufacturability, etc. The series of projects in progress that are aimed at high-speed machines include the development of design and optimization models, control without sensors, magnet materials, prototyping, and testing. An Oak Ridge National Laboratory technical report prepared by Unique Mobility indicates that the cost, volume, and mass goals for the Partnership are realizable with projected design developments and manufacturing improvements.

Novel Power Converters

Current-generation power electronic systems in HEVs include three-phase inverter modules to interface between the dc bus and the electric machine and possibly dc-to-dc converters between the battery and the dc bus. The capacitors at the dc bus and the inductor of the dc-to-dc converter represent major volume and cost elements as do the silicon power devices. Various projects are being conducted that are aimed at reducing the size of the dc-to-dc converter through high-frequency switching, integrated converters that incorporate the battery interface and the machine interface into one converter, and the use of switched capacitor converters, among other things. While progress and success in these projects are expected to lead to incremental improvements in the size and cost goals for the converters, their impact on meeting the goals of the Partnership in terms of performance and cost is unclear.

Integration of Subsystems and Components

The current generation of electrical systems, including electrical propulsion subsystems, power electronic subsystems, and thermal management subsystems, is packaged as discrete and separate elements and assembled together to realize the overall functional objectives. From a systems point of view, integrating these subsystems into a single package and specifying it as a single component can drive their cost and size lower and their reliability higher as the designs continue to evolve. However, the disparate constituent technologies and manufacturing involved in each of these subsystems cause such integration to be a significant challenge. Selected projects are aimed at various degrees of integration: the converter with the motor, the converter package with the thermal management, and so forth. Together, these integration projects are perhaps the riskiest projects in the portfolio. Successful realization of their ambitious goals may lead to significant improvements in the electrical propulsion system performance across various measures.

Summary Assessment

The activities of this program have been multidisciplinary, and its portfolio is diverse: (1) materials development (thermal interfaces, high-temperature insulation, wide band-gap semiconductors, magnetic materials), (2) components development (high-temperature capacitors, silicon on insulator gate drives), (3) converter topologies (multilevel converters, switched capacitor converters), (4) manufacturing (high-temperature packaging), (5) machine control (sensorless operation, optimizing efficiency), and (6) design optimization and modeling (motor design and modeling tools, thermal modeling). Program review reports from 2005 and 2006 indicate a dispersed effort with diverse perspectives (Rogers, 2005; Wall and Rogers, 2006). By contrast, the recent solicitation and its selected projects with specific demonstration goals are aimed at realizing manufacturable engineering prototypes and designs by the selected technology vendors. Overall, these projects are aimed at incremental improvements relative to the state of the art in each of the constituent technologies.

Recommendations

Recommendation. The Partnership should conduct a meta-analysis and develop quantitative models to identify fundamental geometric limitations that ultimately set bounds on and lead to the realization of the size, mass, and cost of power converters and electric propulsion systems in relation to the physical properties of materials and processes such as dielectric strength, magnetic saturation, thermal conductivity, etc. This will allow the various ongoing and future efforts to be benchmarked against the theoretical boundaries of what is possible and enable the establishment of appropriate directions in research goals.

Recommendation. In general, the Partnership should focus on the projects that address specific performance and cost goals of the program on the basis of the results of the meta-analysis recommended above. Specifically, it should (1) intensify packaging efforts; (2) commit additional resources to high-temperature electronics, including wide band-gap semiconductor devices such as SiC; and (3) redirect research on higher speed electrical machines to improve torque density.

STRUCTURAL MATERIALS

Substantial weight savings are a critical and key requirement of the FreedomCAR and Fuel Partnership. The vehicle weight reduction target has been set at

50 percent, with the added criterion of cost parity (i.e., no increase in structural materials cost). This weight target is critical to vehicle design because the powerplant and fuel storage requirements are based on it. Any failure to meet the vehicle weight goal would necessitate a larger powerplant, increased fuel storage capacity, and larger components to achieve the overall vehicle functional goals—for example, driving range and acceleration. As noted in the Phase 1 report (NRC, 2005), the majority of the materials technical programs have been under way for some time, going back to the Partnership for a New Generation of Vehicles (PNGV) before the initiation of the FreedomCAR and Fuel Partnership (NRC, 2001). Thus detailed descriptions of the individual material studies would be superfluous here, and readers interested in specific details are referred to the earlier reports. In this report, the committee concentrates on reviewing the overall objectives and recommending some changes in the program targets and focus that appear to be more appropriate based on the progress to date.

Assessment of the Program

The lightweight materials programs have continued unabated, with only very small modifications to the activities described in the Phase 1 report. Thus the main structural weight savings are anticipated to be achieved through the widespread application of advanced high-strength steels, aluminum alloys (both sheet stock and castings), and cast magnesium. Lesser overall weight savings are also expected from specialized use of glass-fiber-reinforced plastics (GFRPs), titanium alloys, metal matrix composites, and stainless steel. From a technical perspective, all these materials are more than adequately covered in the current programs, from both an application viewpoint and an innovative manufacturing viewpoint.

In addition to the above activities, there is a significant effort on carbon-fiber-reinforced plastic (CFRP) composites. While these materials potentially offer greater weight savings than any of the other candidate materials, the formidable challenges involved in meeting both cost and manufacturing requirements appear to be insurmountable in the time frame of this program, if indeed ever! In particular, the cost barrier is so great that the cost modeling results of the materials technical team apparently confirm the conclusion that CFRP composites will not be economically viable within the time frame of the Partnership. This is not to imply that some research activity in CFRP is not justified, but it does mean that the Partnership should not include such composites as part of its main program objectives and projections for vehicle design.

By far the biggest question mark in the drive to achieve 50 percent weight reduction is the cost penalty for such a massive reduction. There is essentially zero possibility that such a reduction can be achieved at cost parity. However, failure to achieve the 50 percent goal would require redesigning the fuel cell to have a larger capacity; this would be accompanied by a significant increase in the size of the fuel storage system, not to mention an increase in the size and weight of other components such as brakes and suspension. Such enlargements of the fuel cell powerplant and hydrogen storage capability would involve extraordinarily expensive materials and major cost increases, allowing the cost penalty for the structural materials to likely pale by comparison. Thus in the committee's opinion, the 50 percent weight reduction is mandatory even with the associated cost penalty, because the alternative is likely to involve significantly greater cost!

Overall, it seems clear that while the technical challenges to the successful application of lightweight materials are not trivial, a reduction in the associated large cost penalty is by far the bigger obstacle. Minimizing this cost penalty is primarily a matter of reducing the feedstock cost and is influenced to a much lesser extent by component manufacturing improvements, etc. This clearly gets into individual company proprietary matters (e.g., the cost of aluminum sheet stock), so cost modeling of structural materials in the context of the Partnership is unlikely to produce reliable information. However, it is the committee's opinion that some realistic cost estimating to achieve the 50 percent weight reduction is needed if the Partnership is going to obtain anything close to a realistic evaluation of the overall cost/functionality trade-offs.

Recommendations

The technical programs are making good progress, and the team members are extremely competent and interacting very well. The main recommendation of the committee in the Phase 1 report is still pertinent—namely, that the funds supporting materials research should largely be diverted to higher priority areas such as fuel cells, hydrogen storage, high-energy batteries, and infrastructure transitions. The committee believes that the proportion of funding going to materials is still exceedingly high given the urgency of many of the other R&D areas (see Chapter 5). The committee also wants the materials effort to focus on what appears to be the most difficult issue for the materials program: recognition that while a 50 percent weight reduction not only must but also probably can be achieved, it will involve a significant cost penalty but, nonetheless, be the most cost-effective way of achieving the overall vehicle program objectives.

Recommendation. Based on the goal of 50 percent weight reduction as *a critical goal* and the near-certainty that some (probably significant) cost penalty will be associated with it, the Partnership should develop a materials cost model (even if only an approximation) that can be used in a total systems model to spread this penalty in an optimal way across other vehicle components.

Recommendation. The materials research funding should largely be redistributed to areas of higher potential payoff, such as high-energy batteries, fuel cells, hydrogen storage, and infrastructure issues. However, materials research for projects that show a high potential for enabling near-term, low-cost mass reduction should continue to be funded.

REFERENCES

Debe, M. 2007. *Advanced Cathode Catalysts and Supports for PEM Fuel Cells.* Annual Merit Review Proceedings, May 15-18, 2007, Arlington, Virginia. Available on the Web at <http://www.hydrogen.energy.gov/pdfs/review07/fcp_25_debe.pdf>.

Department of Energy (DOE). 2004. *FreedomCAR and Vehicle Technologies Multi-Year Program Plan.* Washington, D.C.: U.S. Department of Energy, Office of Energy Efficiency and Renewable Energy. Available on the Web at <http://www.eere.energy.gov/vehiclesandfuels/resources/fcvt_mypp.shtml>.

Jacobson, D. 2007. *Neutron Imaging Study of the Water Transport in Operating Fuel Cells.* 2007 Annual Merit Review Proceedings, May 15-18, 2007, Arlington, Virginia. Available on the Web at <http://www.hydrogen.energy.gov/pdfs/review07/fc_2_jacobson.pdf>.

James, B.D., and J.A. Kalinoski. 2007. *Mass Production Cost Estimation for Direct Hydrogen PEM Fuel Cell System for Automotive Applications.* 2007 Annual Merit Review Proceedings, May 15-18, 2007, Arlington, Virginia. Available on the Web at <http://www.hydrogen.energy.gov/pdfs/review07/fc_28_james.pdf>.

Kung, H. 2007. *Update on DOE Basic Hydrogen Research.* 2007 Annual Merit Review. Proceedings, May 15-18, 2007, Arlington, Virginia. Available on the Web at <http://www.hydrogen.energy.gov/pdfs/review07/pl_0_kung.pdf>.

Lasher, S. 2007. *Direct Hydrogen PEMFC Manufacturing Cost Estimation for Automotive Applications.* 2007 Annual Merit Review Proceedings, May 15-18, 2007, Arlington, Virginia. Available on the Web at <http://www.hydrogen.energy.gov/pdfs/review07/fc_27_lasher.pdf>.

National Research Council (NRC). 2001. *Review of the Research Program of the Partnership for a New Generation of Vehicles, Seventh Report.* Washington, D.C.: National Academy Press.

NRC. 2005. *Review of the Research Program of the FreedomCAR and Fuel Partnership, First Report.* Washington, D.C.: The National Academies Press.

Rogers, S.A. 2005. *Annual Program Report for the Advanced Power Electronics and Electric Machines Program (November).* Washington, D.C.: U.S. Department of Energy, Office of FreedomCAR and Vehicle Technologies.

Satyapal, S. 2007. *Hydrogen Storage Session Review.* 2007 Annual Merit Review Proceedings, May 15-18, 2007, Arlington, Virginia. Available online at <http://www.hydrogen.energy.gov/pdfs/review07/pl_3_satyapal.pdf>.

Wall, E., and S.A. Rogers. 2006. *Annual Program Report for the Advanced Power Electronics Technology Area.* Washington, D.C.: Department of Energy, Office of FreedomCAR and Vehicle Technologies.

4

Hydrogen Production, Delivery, and Dispensing

PROGRAM OVERVIEW

As discussed in Chapter 1, the FreedomCar and Fuel Partnership in DOE's Office of Energy Efficiency and Renewable Energy (EERE) includes the hydrogen production, delivery, and dispensing program, which is, in turn, part of the Hydrogen, Fuels Cells, and Infrastructure Technologies (HFCIT) program (see Appendix A for an EERE organization chart). This program addresses a variety of means of producing hydrogen, including by biomass gasification and steam reforming of bioderived liquids. The manager of HFCIT is the overall DOE hydrogen program manager. There are three fuel technical teams (see Figure 1-1): fuel pathway integration, hydrogen production, and hydrogen delivery, with participation from DOE and the five energy companies that joined the Partnership 3 years ago. The technical teams report to the Fuels Operations Group, consisting of energy directors and DOE program managers, who in turn report to the Executive Steering Group.

Other activities related to the HFCIT program are in other DOE program offices. The Office of Fossil Energy (FE) supports the development of technologies to produce hydrogen from coal and related carbon sequestration technologies. The Office of Nuclear Energy (NE) supports research on the potential use of nuclear heat to produce hydrogen, and the Office of Science (SC) supports fundamental work on new materials to store hydrogen, catalysts, and fundamental biological or molecular processes for hydrogen production, as well as work potentially affecting other areas of the FreedomCAR and Fuel Partnership. Work on growing, harvesting, transporting, and storing biomass is carried out in EERE but is not part of the Partnership.

TABLE 4-1 Funding Levels for Hydrogen Production, Delivery, and Dispensing Activities in the Partnership

	Funding (millions of dollars)		
DOE Office	FY06	FY07	FY08 Request
EERE/HFCIT[a]	8.4	34.6	40
Fossil Energy (HFI and CCS)	94.9	123.6	91.6
Nuclear Energy[b]	25	19.2	22.6
Science[c]	12.6	13.5	13.7
Total	140.9	190.9	167.9

NOTE: HFI, Hydrogen Fuel Initiative; CCS, carbon capture and sequestration.

[a] The request for FY08 for EERE/HFCIT includes $17 million for work focused on production, delivery, and dispensing in the transition period. Expenditures include conversion of biomass to hydrogen but not growth, harvesting, storage, or transportation of biomass prior to conversion.

[b] Nuclear Hydrogen Initiative; excludes funding for the next-generation nuclear plant (NGNP).

[c] For hydrogen production.

SOURCE: DOE, Answers to questions from committee, pp. 2-9, June 7, 2007.

The budget for the areas relating to hydrogen production, delivery, and dispensing is given in Table 4-1 (DOE, 2007a). In reviewing the hydrogen production, delivery, and dispensing areas, the committee considered whether it is appropriate for the federal government to be involved and, without exception, concluded that these activities are appropriate for federal involvement.

As will be shown in this chapter, DOE through its HFCIT program has made substantial progress on hydrogen production, ensuring that hydrogen can be made available to meet the needs of fuel-cell-powered vehicles as they emerge. However, success in work still under way is needed to minimize cost and to make feasible the production of this hydrogen without increasing carbon dioxide emissions or natural gas imports.

HYDROGEN FUEL PATHWAYS

The hydrogen fuel/vehicle pathway integration effort is charged with looking across the full hydrogen supply chain from well (source) to tank. Specifically, the goals of this integration effort are to (1) analyze issues associated with complete hydrogen production, distribution, and dispensing pathways, (2) provide input to the Partnership on goals for individual components, (3) provide input to the Partnership on needs and gaps in the hydrogen analysis program, and (4) foster full transparency in all analyses. This involves source to vehicle tank analysis, including costs, energy use, safety, and carbon dioxide (CO_2) emissions.

Accomplishment of these goals is overseen by the fuel pathways integration technical team (FPITT) with representation from DOE, the five energy companies, and the National Renewable Energy Laboratory (NREL). FPITT's expertise supports the analysis efforts of the Partnership, coordinates fuel activities with the vehicle systems analysis technical team, recommends additional pathway analyses, provides input from industry on practical considerations, and acts as honest broker for the information generated by other technical teams.

DOE has made important progress toward understanding and preparing for the transition to hydrogen fuel. An effort to develop a transition strategy was established, several workshops to develop scenarios for the transition were held, and a program with three parallel approaches to hydrogen production at fueling stations has been implemented. Program target dates call for completion of the program in about 2017, consistent with the President's goal of enabling large numbers of Americans to choose vehicles powered by hydrogen fuel cells by 2020.

Clearly, there is uncertainty about the time frame in which transitional hydrogen will be required, economically sustainable hydrogen-powered vehicles can be achieved, and a well-developed hydrogen fuel infrastructure can be put in place. Given this uncertainty, the committee believes that DOE needs to incorporate in its studies a time frame for the transition to and subsequent emergence of a mature industry. Thus far, DOE has focused on the transition through 2025, but market sustainability might not be established until 2035 or later. It will take more than a decade to move from the manufacture of a few thousand vehicles per year, when transitional quantities of hydrogen will first be needed, to a mature industry that supports a mature hydrogen production/supply system with centralized production and pipeline distribution. The amount of transitional hydrogen needed over that period will change dramatically. To illustrate this point, the number of hydrogen-fueled vehicles could increase from an assumed 10 million in 2025 to 40 million in the following 10 to 20 if there is a growth rate of 7 to 15 percent annually. Obviously, hydrogen supply would have to grow similarly. Even 40 million vehicles might not be sufficient to stimulate the development of a self-sustaining, mature industry, so transitional methods might be needed eventually to supply even more vehicles and to provide fuel in remote areas. The potential roles of the different transitional hydrogen supply paths need to be viewed from the perspective of this uncertainty. For instance, while transitional hydrogen for 10 million cars might be produced from natural gas without increasing the cost of the natural gas, transitional hydrogen for 40 million cars produced from natural gas would most likely increase the natural gas cost significantly. Thus different energy sources could become important at different points in the transition.

Recommendation. DOE should continue its studies of the transition to hydrogen, extending them to 2030-2035, a transition period during which the number of hydrogen vehicles in use could increase rapidly and use the results of these studies as a basis for evaluating the potential roles of different transitional supplies of hydrogen fuel as demand increases substantially, including both forecourt production at the fueling station and centralized production using the most cost effective means of distributing the hydrogen.

HYDROGEN PRODUCTION

The hydrogen production goals are based on the premise that no single energy source is likely to meet all energy needs in the long term and that U.S. energy security will be enhanced by producing hydrogen from a diverse set of feedstocks.[1] The hydrogen production technical team facilitates the development of commercially viable production technologies. The energy sources under consideration for hydrogen generation, in addition to grid-based electrolysis, are natural gas, coal, biological systems, nuclear heat, wind, and the Sun. Side-by-side comparisons of the cost of producing hydrogen with these different energy resources are not included in this chapter, for two important reasons. First, as described below, the reliability of the estimates varies substantially. Estimates for coal and natural gas are based on actual commercial operating experience, but other estimates, such as those for biomass-derived hydrogen, are based on assumptions yet to be verified. Second, the availability of the resources in this country varies. For example, while the United States has ample supplies of coal for the foreseeable future, natural gas is already being imported to meet current demands. Thus comparison of the different approaches is complex and beyond the scope of this review.

The hydrogen production program includes both long-term hydrogen supply from large, centralized production plants with pipeline distribution and supply during a transition when pipelines will not yet be in place. While it is clear that centralized plants will eventually provide most hydrogen at lower cost than other options, these plants and the necessary pipeline distribution system will not be available initially, when the number of hydrogen-fueled cars in operation will be small, although growing. In addition, once transitional hydrogen supply approaches are in commercial use, it may be economical for the mature industry to continue to rely on them to supply some of the hydrogen needed, particularly in remote areas.

Presently, there is no standard specifying a grade of hydrogen fuel that is acceptable for use in proton exchange membrane (PEM) fuel cell vehicles, and there is no all-inclusive list of maximum acceptable levels of contaminants in hydrogen. While a specification guideline has been issued (Ohi and Hewett, 2005), changes are likely to be made as more data become available. The levels of impurities that are acceptable could significantly affect the cost of hydrogen

---

[1]DOE has calculated that if 300 million vehicles with fuel economy of 60 mpg require hydrogen in 2040, 20 percent of that requirement could be provided by 2 trillion cubic feet (TCF) per year of natural gas, an increase of about 9 percent over the consumption in 2004 (DOE, 2005a). Likewise, producing 20 percent from biomass would require 140 to 280 million metric tons (MMT) (dry) of biomass compared with the 512 to 1,300 MMT currently available potentially from various sources. Similarly, 200 GW of installed wind power would be needed for hydrogen production by electrolysis compared with about 7 GW currently installed; 200 GW of photovoltaics compared with 5,400 MWe currently installed; and 80 GW of nuclear power compared with about 100 GW currently installed (DOE, 2005a).

and the overall efficiency of its production. DOE is well aware of this issue, and efforts are under way to resolve it.

Production Technologies

For centralized plants, the DOE hydrogen production program includes coal gasification, biomass gasification, electrolysis of water with wind energy or off-peak electricity, high-temperature water splitting with nuclear heat, and longer term approaches, including solar electrochemical and biological means. Existing commercial technologies can be used to convert natural gas or coal to hydrogen, and work currently under way at DOE, including the FutureGen program on coal with carbon sequestration, should reduce their costs moderately. Centralized production is visualized for each of these technologies and for natural gas reforming, distributed generation as well. All costs presented here exclude fuel sales taxes.

Hydrogen Production from Coal

Coal is an important potential resource for producing hydrogen because it is cost competitive and relatively abundant in the United States. At current production rates, the nation has over 200 years' supply (see <http://gasprices-usa.com/coal.htm>). Efforts to develop and demonstrate hydrogen production from coal, including coal gasification and CCS, are managed by the DOE's FE. The carbon sequestration subprogram is focused on developing, by 2012, technologies that separate, capture, transport, and sequester carbon, increasing the cost of electricity by less than 10 percent.[2] By that date, the program also plans to have developed a methodology capable for predicting CO_2 storage capacity in a geologic formation to within ±30 percent. This program also has a number of regional partnerships, which include large-scale field tests, site evaluation work, site characterization R&D, collection of information to satisfy National Environmental Policy Act reviews, and other site-related activities to evaluate a variety of geologic formations for sequestration. The technologies developed by the carbon sequestration work will be used to benefit the existing and future fleet of fossil fuel power-generating facilities and provide key technologies and protocols for the FutureGen facility as it looks to capture, transport, store, and monitor the CO_2 injected in geologic formations.

This arrangement—an important part of the program carried out in another program office—presents both management and technology challenges to the Hydrogen Fuel Initiative (HFI) and hence to the Partnership.

This divided responsibility will require close liaison between the managers at

---

[2]For additional information on carbon capture and sequestration technologies and the research program, see <http://www.fossil.energy.gov/programs/sequestration/capture/>.

HFCIT and FE. In the Phase 1 report, the committee recommended strengthening this liaison. DOE concurred and has improved its communication and coordination by taking a number of actions:

- Setting up a hydrogen coordinating group composed of representatives from EERE, FE, NE, Basic Energy Sciences (BES), and the Department of Transportation (DOT);
- Establishing an interagency working group to address hydrogen coordination issues among federal agencies; and
- Using its systems analysis capabilities to illuminate the implications for the Partnership of any cost and schedule issues that might arise in the FE program.

The committee appreciates the value of these actions but notes that such mechanisms add value only insofar as they are used. It urges continued attention to building a highly effective liaison through these coordination arrangements. It believes as follows:

- That CCS will pace the use of coal to produce hydrogen, and
- That the technical and economic feasibility of capturing by-product CO_2 and shipping it to permanent underground storage while producing electricity and hydrogen from coal will have to be demonstrated before significant commercial investment can materialize.

A demonstration is being planned through FutureGen, a 275-MW, $1.8 billion integrated gasification combined cycle (IGCC) plant that will gasify the coal to produce electricity and hydrogen and sequester the resulting CO_2. The committee believes that the general technology and system concepts embodied in FutureGen now offer the most promising way to produce hydrogen from coal while minimizing CO_2 release. And since FutureGen is now the principal platform for demonstrating a practical, commercial CCS system, its implementation will determine when the large-scale production of hydrogen from coal is introduced.

To the extent that this project is delayed or fails to provide evidence acceptable to the public that CCS affords adequate protection at an acceptable cost, hydrogen production from coal will suffer corresponding delays. These delays could have multiple causes—for example,

- Simple slippage in the FutureGen project schedule, a possible consequence of underfunding, unforeseen technical problems, siting difficulties, and so forth,
- Issues arising from ambiguity surrounding regulatory authority,
- Liability concerns,
- The inability of FutureGen to provide a model for the commercial de-

velopment of CCS, much as the Power Reactor Demonstration Program of the 1950s failed to provide such a model for nuclear energy, or
- Other difficulties, unforeseeable now but arising over the course of the project.

Whatever their source, delays to CCS pose a risk to the Partnership and HFCIT goals for hydrogen production from coal.

Recommendation. DOE should conduct a systematic review of the CCS program as it affects the schedule for and program assumptions about hydrogen production from coal. This review should identify indicators of incipient program slippage and, through systems analysis, the program consequences of possible delays, leading to recommendations for management actions that would compensate for these delays.

Hydrogen Production from Nuclear Heat

NE seeks to demonstrate by 2017 the commercial-scale production of hydrogen using heat from a nuclear energy system. Some advanced nuclear reactor designs operate at very high temperatures, making them well suited for thermally driven hydrogen production processes. These high-temperature reactors remain in early development by the Generation IV Nuclear Energy Systems Initiative (Generation IV) and could provide the low-cost heat necessary to produce low-cost hydrogen.

The nuclear hydrogen program is managed under three technology thrusts:[3]

- *Thermochemical water-splitting cycles.* Thermochemical cycles convert water to hydrogen and oxygen using chemical catalysts at high temperatures. These processes have the potential for high-efficiency hydrogen production on a large scale, but the technology remains in a very early stage.
- *High-temperature electrolysis.* Also called steam electrolysis, this technology uses electricity to produce hydrogen from steam instead of from liquid water. It promises higher efficiencies than standard electrolysis, which might be used at the forecourt of fueling stations during a hydrogen transition. This, too, is in an early stage, and the chief technical challenges include the development of high-temperature materials and membranes.
- *Reactor/hydrogen process interface.* The interface between the nuclear reactor and the hydrogen production system presents severe challenges to any working system—long heat-transfer paths at elevated temperatures;

[3]The program is summarized at <http://www.hydrogen.energy.gov/nuclear.html>.

heat exchangers that are subject to elevated temperature and a corrosive chemical environment; new safety and regulatory issues; and support systems for chemical processes and hydrogen and oxygen storage. In addition, systems studies seek to focus this complex program and improve coordination.

In all of these research areas, much basic work must be completed before a development and demonstration program can be properly contemplated. Nevertheless, nuclear production of hydrogen remains an important option—it is potentially lower in cost and could be a hedge against delays in CCS technologies or against coal shortages.

Recommendation. Like the hydrogen production from coal option, the Hydrogen, Fuel Cell and Infrastructure Technology (HFCIT) program should actively employ the liaison mechanisms put in place since the Phase 1 review. However, the exploratory nature of the programs for nuclear production suggests that, unlike the coal option, a detailed systems analysis of schedule delays would be premature at this time. Instead, systems analyses should focus on the complex interactions among program components, especially between the research elements of the nuclear and chemical processes, to ensure that technical progress in each distinct area leads ultimately to a practical system.

Hydrogen from Electrolysis

The electrolysis of water, though energy intensive, is one of the few options under consideration for distributed, on-site, point-of-use production and delivery of hydrogen when conventional sources and processes are not available. When coupled with a renewable power generation scheme such as for wind or solar power, the overall advantages are considerable, especially when carbon emissions are taken into account. However, the relative siting of the power generation and electrolysis devices is an issue since a location suitable for, say, a wind farm might not be suitable for the hydrogen generator. Centralized electrolysis processes are also under development to reduce operating and capital costs. DOE recognizes the importance of electrolysis and has been engaged in facilitating new concepts, advances, demonstrations, and analyses. Clearer insight into the challenges of the program cost targets—for distributed generation, \$3.70/gallon of gasoline equivalent (gge) in 2012 and <\$3.00/gge in 2017, assuming grid electricity costs \$0.05/kWh and units that produce 1,500 kg H_2/day;[4] for centralized generation with wind energy, \$3.10/gge in 2012 and <\$2.00/gge in 2017, excluding delivery

[4]DOE, Hydrogen Production Technical Team, Presentation (Slide 27) to the committee on March 2, 2007.

costs—has been gained from the recent analysis efforts. It is still too early to predict the probability of meeting the longer-term targets.

The budgets for electrolysis R&D have been increasing ($1.6 million in FY05, $3.5 million in FY07). Although funding has more than doubled, it is not enough, as will be discussed. Conventional water electrolysis technology is more mature than other processes, in part because extensive operating knowledge, systems design and engineering, and applications have been in place for decades, specifically in military applications (submarines). Large commercial processes have also been available. As a result, proof-of-concept programs are not warranted, but there are significant matters—cost, systems integration, analyses, and field trial results—that need to be better understood. The NREL wind-electrolysis demonstration (Harrison, 2007), as well as its source-to-wheels analyses, has made a significant contribution, and its results have led to the refinement of targets.

The technology is challenged primarily by costs, in particular the cost of electrical power to split the water. The fundamental energy requirements for this process will not change, but overall system costs are addressable. Sensitivity and trade-off analyses will be needed to delineate the most attractive scenarios, in part because two distinctly different technologies (membrane and alkaline) are under consideration. Because power requirements are nearly fixed, the potential for capital cost reduction for each technology will be an important outcome of these sensitivity analyses. Both technologies have extensive histories: Membrane electrolysis offers the advantage of high hydrogen generation rates and efficiencies and the promise of further enhancing these rates (thereby reducing capital outlay), whereas alkaline systems have track records for lifetime, reliability, and lower capital costs. Both technologies have demonstrated high-pressure generation capability, and both lend themselves to minimizing downstream cleanup, storage, and compression. It is too early to predict the outcome of solid oxide electrolysis.

There are approximately eight funded PEM, alkaline, and solid oxide electrolysis projects, as reported by General Electric at the DOE 2007 merit reviews.[5] These projects are engaged in analyses, component development, and demonstrations. Additional ongoing research looking at longer-term possibilities such as the photoelectrolysis of water is basic in nature. The efforts are aimed at reducing the capital costs of the hardware and the number of parts and at finding new membranes, catalysts, and materials of construction. Although such initiatives are appropriate, the costs of electrolysis will always be dictated by the power requirements for splitting water, which limits what is achievable by reducing hardware costs. To date, the mass manufacture of electrolyzers has not been addressed in the commercial sector because the market is still small. Once manufacturing ac-

[5]R. Bourgeois, GE, "Advanced alkaline electrolysis," Presentation to the DOE 2007 Annual Merit Review in May 2007. Available on the Web at <http://www.hydrogen.energy.gov/pdfs/review07/pdp_16_bourgeois.pdf>.

tivities begin to take off, they will probably contribute significantly to reducing costs. In addition, the considerable overlap between membrane electrolyzers and membrane fuel cell components enhances supply chain strengths and capabilities and facilitates the development of new materials (e.g., membranes).

Recommendation. The DOE should continue to promote electrolysis that uses renewable power integrated with electrolysis systems and to support analyses and demonstrations. High-temperature electrolysis activities within the Office of Nuclear Energy should be closely monitored.

Recommendation. The Partnership should increase funding for electrolysis programs to advance the technology, demonstrations, and systems integration.

Recommendation. Basic Energy Sciences should support, as appropriate, fundamental research in the area of catalysts, membranes, coatings, and new concepts.

As mentioned earlier, a population center where distributed hydrogen production would be needed is not usually in a place where significant electricity is generated from the wind or the Sun. As a result, it is not clear, based on current understanding, how a generator of power from the wind should be situated relative to a generator of hydrogen to maximize the benefits. For example, depending on the specifics of the location, it might be more efficient to generate electricity in a wind farm and transport it over the grid to distributed electrolyzers than to cogenerate power and hydrogen and transport the hydrogen to the fueling site. Likewise, the extent to which wind power could be generated at the site of a distributed hydrogen generator is unclear.

Recommendation. DOE should undertake a systems study to determine how best to use wind power–electrolysis combinations to generate hydrogen, considering overall cost and efficiency.

Hydrogen Production from Biomass

Biomass is a renewable and potentially sustainable source of liquid fuels and hydrogen. A comprehensive study jointly sponsored by DOE and the Department of Agriculture (DOE, 2005b) concluded that the United States has sufficient land resources to sustain production of biomass to supply 30 percent or more of the nation's current consumption of liquid transportation fuel. To achieve that level of production, it is assumed that three times more forest biomass than today will be collected; that crop yield will be increased by 50 percent and recovery of crop residues by 75 percent; that 55 million acres will be dedicated to the production of perennial bioenergy crops; and that other non-farm-use residues will be converted to biofuels. In addition, a cost-effective and energy-efficient process to convert cellulosic material to biofuel will be needed, which requires innovation at various stages, including crop production and biomass degradation.

DOE has set goals to supply 20 percent of liquid transportation fuel by 2017 and 30 percent by 2030. These are ambitious goals, and DOE is investing substantial resources in R&D to achieve them, including a FY08 request for $375 million. The FreedomCAR and Fuel Partnership includes only the technology for conversion of biomass to hydrogen. The committee did not attempt a comprehensive review of the program and budget on biomass growth, harvesting, collection, and transportation to conversion plants. However, it is clear that there are significant hurdles to overcome, and it will be a stretch to achieve these goals (DOE, 2006a). Work to date has not established how much can be recovered sustainably at target costs and by target dates. To make this estimate involves resolving various technical barriers that DOE has identified relating to biomass availability and cost (DOE, 2003), as well as land and water use issues and competition for both resources. The committee believes that the impact of biomass on the supply of hydrogen cannot be reliably estimated until programs relating to biomass production, harvesting, collection, storage, preprocessing, and transportation can define commercially viable pathways from crops or other biomass sources to hydrogen production plants and until the specific government-sponsored incentives become clear, along with land and water use policies that may be required to stimulate wider use of this option. Resolving these issues will require the involvement of other government departments, including the Department of Agriculture. The committee believes that early definition of government-sponsored commercial incentives and land and water use policies would help facilitate the later development of appropriate government actions in these important areas.

Nonetheless, the Partnership is anticipating the possibility of a significant increase in the use of bioethanol in transportation fuel (say, E85 or E15) in the near future, from about 1.7 percent today (DOE, 2006b), and the potential impact on combustion engine technology of such an increase. In response to this, DOE initiated solicitation DE-PS26-07NT43103, which includes development of engines for flexible-fuel, light-duty vehicles (FFVs) optimized for operation on ethanol-gasoline blends up to 85 percent ethanol by volume. This is in line with Recommendation 3-3 of the Phase 1 report.

Hydrogen can be produced by gasifying the biomass feedstock directly or by gasifying one of the conversion products, such as ethanol. These conversion technologies are known, but the current cost of production is high, $7.00/kg hydrogen ($H_2$) at a feedstock cost of $53/ton.[6] DOE projects that a lower feedstock cost, more energy-efficient production, and cost reduction due to a larger scale of operation will result in a cost of $3.50/kg H_2. The committee believes that $3.50/kg is

[6]These estimates are taken from *The Hydrogen Economy* and are for hydrogen delivered by tank trucks from a high-pressure oxygen gasifer producing 24,000 kg H_2/day (NRC/NAE, 2004).

a stretch goal that could be achieved only after overcoming significant technical and policy hurdles, as described earlier. Thus the extent to which biomass will become a source of hydrogen is highly uncertain, and DOE should continue to investigate a broad portfolio of hydrogen production technologies.

As part of the effort to reduce gasification costs, DOE is now considering a 155,000 kg H_2/day Battelle biomass gasification plant as opposed to the 24,000 kg H_2/day high-pressure gasifier that was assumed in *The Hydrogen Economy* (NRC/NAE, 2004). Neither operation has yet been fully demonstrated. The much larger plant needs 2,425 tons of biomass/day with current technology and 2,125 tons/day with projected future technology compared with 442 tons/day biomass for current technology and the smaller plant. The larger plant needs 307 square miles to support it with current technology and 180 square miles with future technology. The number of sites that could support such a large biomass plant and still have acceptable delivered biomass cost and delivered hydrogen cost needs to be determined. DOE estimates there are roughly 50 potential sites throughout the country with current biomass yields and upwards of 100 sites that could yield biomass with future crop technology. This needs to be confirmed.

Biomass gasification technology is promising, but much remains to be done to put it on a solid basis. How the different types of biomass feedstock must be prepared for ease of delivery and reliable processing in the gasifier needs to be determined. Bench-scale, pilot plant, and semicommercial-scale work are needed to have a firm basis for scale-up to a 2,125 tons/day plant or larger. Also, gas cleanup and separation into pure hydrogen needs to be demonstrated while dealing with contaminants and tar. The committee judges that it will be difficult to achieve this technology by 2017, and several years more may be needed. However, if successful, hydrogen supply from biomass gasification could supplement other sources of hydrogen, and the committee continues to believe that this program is a very important element in the portfolio of hydrogen production technologies.

The impact of biomass on future hydrogen supply is difficult to evaluate, in part because there are so many alternative paths. In addition to gasification of biomass to hydrogen there are other pathways that may be reasonable. For instance, gasifying biomass to make an alcohol mixture and then reforming the alcohol to hydrogen at distributed locations would eliminate the need to distribute hydrogen itself. Distributed reforming of cellulosic ethanol, aqueous glucose, or aqueous lignins is another possibility. Additionally, biological methods discussed in *The Hydrogen Economy* (NRC/NAE, 2004) but still in the basic science phase, could one day be significant. Finally, cofeeding biomass with coal in a coal gasification process might eventually be attractive. With the current state of knowledge, the committee believes it is not yet possible to identify the preferred approaches and encourages DOE to focus its program efforts on studies that will enable this identification.

The committee notes that while DOE has involved the agribusiness partners in the biomass program, the energy partners are involved primarily in the conversion or gasification step and not in the harvesting and so on of the biomass feedstock. The committee believes that it would be important to bring the commercial perspective of the energy partners to all aspects of the biomass program, since these partners have long experience with creating and supplying transportation fuels from natural resources and have been involved in converting coal and other solid energy sources to transportation fuels.

Recommendation. The committee recommends that DOE projections of future hydrogen production for hydrogen-powered vehicles include scenarios in which the timetable for commercial quantities of these fuels is delayed, perhaps by as much as a decade.

Recommendation. DOE should give priority to completing process development on biomass gasification, including any needed demonstration projects.

Recommendation. DOE should undertake a systems study to assess the relative importance of barriers to biomass production, availability, transportation, and conversion to hydrogen; to identify the areas that are most important to commercial viability; and to give them priority. This study should address technical barriers already identified, including impact on the environment, and help define policies for land and water use and government-sponsored commercial incentives that would stimulate commercial expansion of the biomass options.

Recommendation. DOE should involve the energy partners in all biomass programs related to conversion to hydrogen or hydrogen carriers as quickly as possible.

Recommendation. Given the large number of potential ways of using biomass to supply hydrogen, DOE should identify the most promising approaches so it can focus on options that could have the greatest impact on hydrogen supply.

Special Production Considerations During Market Transition

No one knows just how hydrogen will be supplied during the transition period when fuel-cell-powered cars first become available. However, it is reasonable to expect that it will first be supplied to fueling stations from existing centralized sources and distributed as a gas by tube trailer or as a liquid by carrier, since a pipeline distribution system similar to the system for natural gas will not initially be available. These supplies are likely to be supplemented increasingly with time, by facilities at forecourts to produce hydrogen locally, with no need for distribution, using steam reforming of natural gas from the existing supply system or electrolysis with electricity from the grid. As the demand for hydrogen grows, its distribution by pipeline from centralized sources to forecourts will become economically attractive in highly populated areas, and it will be used more and more. However, in some remote areas such a pipeline system may never become economical, owing to low demand, and the sources of fuel used at the beginning of the transition could continue to be used.

Three approaches to the forecourt generation of hydrogen are being studied. They involve the reforming of natural gas taking advantage of the existing natural gas distribution system; the reforming of ethanol or other bioderived liquids; and the electrolysis of water. Dehydrogenation of a carrier liquid that is subsequently returned to a refinery or chemical plant for rehydrogenation is another option.

DOE has made substantial progress in understanding the transition to a sustainable market, defining requirements for forecourt production systems based on natural gas reforming that meet initial cost targets and advancing other options for onsite generation. Natural gas reforming is well-established technology, and program efforts have been directed toward the specific requirements of practicing it at fueling station sites in relatively small units (e.g., 1,500 kg H_2/day). DOE has established that hydrogen can be produced from natural gas in an integrated system for the target cost of $3.00 per gallon gasoline equivalent (gge)[7] and has a cost goal of $2.50/gge by 2010. A study carried out for DOE has estimated that the natural gas required in 2025 to fuel 11.6 million hydrogen vehicles in the 27 largest U.S. cities would increase gas demand on average by 2.1 percent over the demand in 2004, although this percentage would vary from city to city, and the cost of additional gas transmission lines to transport that gas might cost $1.0 billion to $1.5 billion (Energy and Environmental Analysis Inc., 2006). There is much uncertainty regarding the effect of small increases in natural gas demand on price, but higher demand would likely increase natural gas imports. Given that 11.6 million vehicles would be only about 15 percent of the census vehicle population in those 27 cities, natural gas price and supply could become an important issue as the penetration of hydrogen cars increases in those 27 cities and expands into smaller communities, assuming all the hydrogen is made from natural gas. The foregoing discussion highlights the importance of having other ways of making hydrogen for the transition.

Important issues remain in the design of reformer systems for forecourt use. These issues could substantially constrain all approaches to distributed hydrogen generation at forecourts, probably limiting hydrogen availability in the early years of its commercial introduction. The current design requires 6,500-7,000 square feet of space, and local regulations will require additional setbacks from adjacent structures. Based on a sample of 120 existing automobile fueling stations in New York, Los Angeles, and Dallas, DOE concluded that it would not be feasible to place onsite reforming in 40 percent of them (49 stations) and would

[7]Assumes the manufacture of 500 reformer units per year, each with a production of 1,500 kg H_2/day. This assumes natural gas at $5 per million Btu, a capacity factor of 70 percent, and a production efficiency (energy content of H_2/input energy content of natural gas) of 69 percent.

be clearly feasible in only 13 percent (16 stations).[8] The committee believes that the DOE estimate of 16 clearly feasible sites is optimistic because of the inherent difficulties of retrofitting an existing site with new facilities and that an even lower percentage of sites would, in fact, be feasible with present design and safety considerations. DOE is aware of the importance of reducing the space requirement and is considering alternative designs, including smaller reformers and less hydrogen storage. Because the budget request for FY08 includes no funds for distributed natural gas reforming, additional funding will be needed to complete this important work.

The committee believes that the target cost of $2.50/gge for distributed generation is very optimistic based on current technological routes, particularly in view of the need to reduce the size of the generator to minimize forecourt space requirements. It believes, accordingly, that DOE needs to reevaluate this target taking into consideration the constraints and approaches available for improvement as well as the latest gasoline price outlook.

Recommendation. DOE should put more emphasis on the space requirements for forecourt hydrogen generation by studying ways to minimize these requirements.

HYDROGEN DELIVERY, DISPENSING, AND TRANSITION SUPPLY

Overview

Unlike the traditional petroleum delivery system, whereby gasoline and diesel fuel are delivered from refineries to fueling stations and stored there at relatively low cost and low energy consumption, the system for delivering hydrogen from central production to a refueling station with subsequent storage and dispensing to vehicles will be a significant factor in hydrogen fueling. Similarly, in a fully developed hydrogen economy, delivery, storage, and dispensing at a high pressure will probably cost as much as production and will consume more energy. Distribution costs are even more of a concern during the early transition years, when there is a lack of hydrogen demand, particularly where central production will be the source of hydrogen. Identifying central hydrogen supply during the transition, whether excess capacity or dedicated supply, could provide significant opportunities to ease the transition to a hydrogen economy.

It is likely that some of the needed hydrogen would be supplied from existing facilities. To achieve the lower sulfur levels of conventional fuels (e.g., gasoline and diesel) when refining heavier crude oils, the industry will have to increase its hydrotreating capacity from 14 million barrels per day in 2004 to over 27 million

[8]DOE, Answers to questions from committee, p. 59, received April 17, 2007.

in 2030.[9] At least part of this hydrogen will be generated by gas supply companies for over-the-fence sales to refineries, and these gas companies, which already produce and deliver gaseous and liquid hydrogen, are likely to see hydrogen vehicles as an attractive additional market.

There are five main ways to deliver hydrogen from central supply to refueling stations:

- *Pipeline delivery.* This requires storage at the production site, laying pipeline to the forecourt, and onsite storage as a gas at the forecourt. It is the lowest cost route and utilizes pipeline delivery technology that is well known and has been in commercial use for decades. Energy losses with a pipeline are less than those with the other delivery methods (see Table 4-2), but pipeline delivery probably will take the longest time to implement because of permitting and rights of way. Production and forecourt storage costs should be lowest of all supply modes since the pipeline itself is the vessel that stores the surge capacity.
- *Liquid delivery.* This involves liquefaction at the production site, delivery of the liquid by truck to the forecourt, liquid storage at the refueling site, and dispensing as a high-pressure gas to the vehicle after vaporization and compression. Liquefaction increases the cost of hydrogen production and decreases its efficiency by approximately 16 percent (see Table 4-2). However, the high density of the liquid allows the highest payload by weight of hydrogen to be moved by truck. Furthermore, storage at the site of high-pressure gas can be a significant forecourt issue because of the space required, and liquid minimizes this. Since the basic technology has been practiced for a long time, only incremental improvements are expected in liquefaction and distribution, and these costs will probably continue to be about 50 percent more than pipeline costs.
- *Gaseous delivery.* This requires high-pressure gas storage at the production site, delivery by high-pressure tube trailers, and high-pressure gas storage at the forecourt. Currently, the lower delivery density makes this option impractical for stations that experience high demand for hydrogen. As a result, it takes 12 to 15 high-pressure tube trailers to deliver the same payload as one liquid hydrogen trailer. In addition, because today's gaseous delivery technology costs several times as much as the technology for pipeline delivery depending on volume, it may not be cost-effective. New technology developments focus on increasing the payload using cryogenic gaseous hydrogen storage—that is, storage as a gas at low temperatures and/or higher pressures.
- *One-way liquid carrier delivery.* This requires methanol, ethanol, or a similar hydrocarbon and would require reforming at the forecourt to

[9]DOE, Answers to questions from committee, p. 31, received April 17, 2007.

TABLE 4-2 Delivery and Dispensing Energy Efficiency[a]

	Efficiency (%)		
Fuel/Delivery Mode	Well (Source) to Tank	Tank to Wheels	Well (Source) to Wheels Overall
Gasoline	81	17	14
H_2/pipeline	64	41	26
Liquefaction	48	41	20

[a]Calculations are based on the Argonne National Laboratory's Greenhouse Gases, Regulated Emissions, and Energy Use in Transportation (GREET) model. For analyses of a variety of fuels, see results at <http://www.transportation.anl.gov/pdfs/TA/273.pdf>.

generate hydrogen. Commercial methanol processes exist, and ethanol reforming processes are being developed. While methanol and ethanol reforming onsite may be more expensive than natural gas reforming, they could have some advantages (costs and efficiency) from a delivery standpoint, particularly during the transition. The higher cost of reforming at the forecourt could more than offset the high costs for liquid or gaseous hydrogen delivery, on a source-to-tank basis. For example, ethanol is a dense liquid containing 12 percent hydrogen that can probably be transported with existing infrastructure.

- *Two-way liquid carrier delivery.* Such liquid carriers could be hydrogenated to more than 13 percent hydrogen content, transported to the forecourt, dehydrogenated, and returned to the hydrogenation site for rehydrogenation. With current technology, they can be hydrogenated to 7-8 percent hydrogen. This is a long-term option that requires R&D and systems analysis to develop an understanding of energy and cost issues. For example, if the liquid could also serve as onboard storage for hydrogen, the overall system would look something like the gasoline system but would increase vehicle complexity and cost. With delivery and dehydrogenation to only the forecourt, there could be some advantages, particularly where eventual pipeline delivery might be difficult or costly.

The DOE Program

DOE's plan for delivery, storage, and dispensing is robust and was developed with aggressive cost targets (Table 4-3). The goal is to reduce the cost of delivery plus dispensing to less than $1.00 per kilogram hydrogen by 2017. This compares to the current costs of \$3-\$4/kg H_2 at low volume and \$2-\$3/kg H_2 at high volume. Given that hydrogen pipeline, truck delivery, compression, and storage technologies have been practiced for decades by the gas industry, the committee questions whether it will be possible to reduce costs by a factor of 2 or 3.

TABLE 4-3 Cost Targets for Hydrogen Delivery and Dispensing (dollars per kilogram of hydrogen)

Activity	2010	2012	2015	2017
Delivery from central plant to refueling gate		<0.90		<0.60
Dispensing at refueling site[a]	<0.80		<0.40	

[a]Includes compression and storage.
SOURCE: Based on J. Kegerreis and M. Paster, DOE, delivery technical team, Presentation to the committee on March 1, 2007.

The program for delivery, storage, and dispensing has been slow to start up, especially on the delivery side, due to congressionally directed funding in the overall HFI. This very important program has been consistently underfunded since HFCIT started, with $16.9 million budgeted but only $8.3 million funded in the 2004-2007 period. It appears that the DOE is working well with gas companies and the Partnership's delivery technical team (DTT). While the program is at risk as a result of past underfunding, some very important analysis work has been accomplished. The development of the Lighthouse concept for market penetration has been a significant accomplishment and helps the DTT team focus on specific delivery, dispensing, and early supply options and issues. Also, the completion of the H2A model, whose components and submodels define delivery and dispensing, is significant and will help to better delineate and evaluate scenarios for getting hydrogen to fuel cell vehicles during the transition and later. The H2A model has also played a significant role in developing the research plan. Finally, analytical work on forecourt issues has progressed—for instance, the very excellent analysis of overall U.S. natural gas supply and demand and of the issues involved in getting natural gas to the refueling station for on-site steam methane reforming (SMR).

The future program is built around the DTT Roadmap, which identifies the following key challenges:

- *Pipelines.* Metal embrittlement, capital cost reduction, urban distribution issues, composite materials for construction, use of existing natural gas pipelines.
- *Compression.* Reliability/durability, new technologies, and the energy efficiency and size of the refueling station.
- *Liquefaction.* Dramatic cost reduction, dramatic increase in energy efficiency, boil-off.
- *Off-board storage.* Lower capital costs, cryogas, other hydrogen carriers, suitability of geologic storage.

TABLE 4-4 Budgets for Hydrogen Delivery Activities (millions of dollars)

	FY04	FY05	FY06	FY07	FY08
Budget request	1.0	3.8	5.9	6.2	8.0
Pipelines				1.8	
Compression				0.7	
Storage				0.8	
Liquefaction				1.2	
Carriers				1.0	
Analysis				0.7	
Expenditures	0.4	2.8	1.1	4.0 (spend rate)	

SOURCE: J. Kegerreis and M. Paster, DOE, "Hydrogen delivery," Presentation to the committee on March 2, 2007.

- *Gaseous tube trailers.* Increase capacity fourfold with higher pressure, cryogas, or other hydrogen carriers.
- *Carriers.* Liquid one-way and two-way carriers; solid carriers.

The DOE budget for the program is shown in Table 4-4. Overall, a lot has been accomplished, but much more progress is needed. The Partnership's DTT has identified important R&D areas for improving the cost and energy efficiency of the delivery and dispensing of hydrogen. However, in light of how important delivery is to both the market transition and sustained market penetration time frames, it deserved more funding and attention. The program is most likely underfunded even at the FY08 $8 million level requested.

Recommendation. DOE should increase funding for the delivery and dispensing program to meet the market transition and sustained market penetration time frames. If DOE concludes that a funding increase is not feasible, the program should be focused on the pipeline, liquefaction, and compression programs, where a successful if only incremental short-term impact, could be significant for the market transition period.

Recommendation. DOE should, with the guidance of an independent outside committee, evaluate the achievability of the program's 2012 delivery and dispensing cost goal, \$1.00/kg H_2, particularly with 700 bar (10,000 psi) gas dispensing.

Home Refueling

One path that could reduce or even eliminate the need for a hydrogen distribution and delivery infrastructure is home refueling, which would allow consumers to refuel their vehicles at home. Furthermore, if additional benefits could be attained with such a device, then the value would be much greater—if, for example, the system could provide both onsite heat and power. So far, two approaches have been proposed for refueling hydrogen vehicles at home: (1) integrating the hydrogen generation and delivery operations with a stationary fuel cell system that generates electricity for the home and is fueled by a natural gas (propane) or by a liquefied propane gas (LPG) reformer or (2) a home-based water electrolysis unit.

To demonstrate these approaches to delivering hydrogen to fuel cells, Honda (<http://world.honda.com/FuelCell/FCX/station/>) has units running in California and in New York, while GM (<http://www.usatoday.com/money/autos/2006-09-24-gm-hydrogen-usat_x.htm>) and others have announced plans for similar devices in the years to come. To provide fuel to the car, a slipstream of the hydrogen used to generate home electricity is diverted to downstream purification and compression stages, followed by storage and dispensing to the vehicle. The fuel cell brings the added value of distributed heat and electrical power to the site while enabling the generation, storage, and dispensing of hydrogen at the convenience of the home owner. From a technology perspective, many of the hydrogen subsystems of the devices (generation, delivery, storage) are based on the same technologies under development in other programs currently funded by DOE, including the fuel cell itself.

Water electrolysis home refueling is based on a conventional electrolytic process followed by purification, compression, and storage stages. Although apparently simpler from a hydrogen generation perspective, the limitations of the water electrolysis process (it requires relatively high investment and quantities of electricity) must be taken into account along with the challenges of purification, compression, and storage, as in the previous case. One more factor to be considered in this case is the cost of power to electrolyze water. As in the aforementioned case of fuel cells, DOE is supporting selected aspects of electrolysis technologies.

Regardless of the path chosen—electrolysis or reformation/fuel cell—the home refueling device does not require additional component development initiatives, as it will continue to benefit from advancements made in such areas under the existing efforts. Additional funding is therefore not required from this perspective. Integration, demonstrations, and siting will need to be addressed as will safety, permitting, and codes and standards of a hydrogen-based process in residential environments.

Recommendation. DOE should consider supporting advanced systems engineering, integration, and demonstrations for home-based refueling systems, which should bring substantial learning value for such systems. This program should include careful consideration of operation and maintenance procedures that home owners are willing and able to perform.

REFERENCES

DOE (Department of Energy). 2003. *Roadmap for Agricultural Biomass Feedstock Supply in the United States*, DOE/NE-ID-11129, Revision 1, November 2003.

DOE. 2005a. *Multi-Year Research, Development and Demonstration Plan: Planned Program Activities for 2005-2015.* Section 2.3, Domestic Resources for Hydrogen Production. Washington, D.C.: Office of Energy Efficiency and Renewable Energy. Available on the Web at <http://www1.eere.energy.gov/hydrogenandfuelcells/mypp/>.

DOE. 2005b. *Biomass as Feedstock for a Bioenergy and Bioproducts Industry: The Technical Feasibility of a Billion-Ton Annual Supply.* April, DOE/GO-102005-2135, Robert D. Perlack, Lynn L. Wright, Anthony F. Turhollow, Robin L. Graham, Bryce J. Stokes, and Donald C. Erbach.

DOE. 2006a. *Breaking the Biological Barriers to Cellulosic Ethanol: A Joint Research Agenda.* DOE/SC-0095.

DOE. 2006b. *Annual Energy Outlook 2006.* Washington, D.C.: Energy Information Administration.

Energy and Environmental Analysis, Inc. 2006. *Initial Look at Potential Natural Gas Infrastructure Constraints Related to Transition to Hydrogen Transportation Fuels.* September 13.

Harrison, K. 2007. *Renewable Electrolysis Integrated System Development and Testing.* Annual Merit Review Proceedings. Arlington, Virginia, May. Available on the Web at <http://www.hydrogen.energy.gov/pdfs/review07/pd_8_harrison.pdf>.

Ohi, J.M., and R. Hewett. 2005. *The Department of Energy's Hydrogen Safety, Codes and Standards Program: Status Report on the National Templates.* Society of Automotive Engineers Paper SAE J2719, November.

National Research Council/National Academy of Engineering (NRC/NAE). 2004. *The Hydrogen Economy: Opportunities, Costs, Barriers, and R&D Needs.* Washington, D.C.: The National Academies Press.

5

Overall Assessment

MAJOR ACHIEVEMENTS AND TECHNICAL BARRIERS

When the section corresponding to this one was written for the National Research Council (NRC) Phase 1 review, the achievements appropriately included several nontechnical components of the Partnership, among them the then-new elements of planning and organization that promised to help the program accomplish its major goals. These elements are all now in place and are providing the positive results that were expected. They are important overall program achievements but will not be revisited.

Even though some of the achievements are outcomes of earlier work, emphasis is placed on accomplishments considered to be especially noteworthy and that were conducted since the Phase 1 review. The following is a brief summary of achievements and remaining barriers in several key areas.

Advanced Combustion and Emission Control

Since advanced internal combustion engine (ICE) hybrid and plug-in hybrid vehicles can provide significant petroleum savings and emission reductions during the transition to a more hydrogen-dominated transportation scenario, technology advancements leading to improvements in ICE efficiencies as well as reduced tailpipe emissions are very important to the Partnership. Accomplishments include the following:

- A continued deepening of the fundamental understanding of the governing thermochemical processes that control alternative advanced combustion and aftertreatment operation. Research efforts are now being guided through fundamental analyses based on laboratory measurements supplemented with advanced simulation.

- The establishment of the working group Crosscut Lean Exhaust Emissions Reduction Simulation (CLEERS), whose membership of industry, academic, and government researchers collaborates to guide research activities.
- Demonstrated peak thermal efficiency of laboratory engines operating at speeds and loads corresponding to peak efficiency has increased about 2 percentage points to over 41 percent. This represents an increase of about 10 percentage points compared to current OEM engines.
- Experimental demonstration of Bin 5 emissions using a NO_x adsorber and a urea selective catalytic reduction (SCR) system.

The technical barriers for the advanced combustion and emissions control technologies are those of implementation, development, and cost. Specifically,

- Implementation and control of advanced combustion approaches into the operating regime of the engine, which includes combustion mode switching and transients.
- Developing the aftertreatment systems that will effectively couple with exhaust gas characteristics of advanced combustion approaches and fuel changes.
- Reducing the cost of aftertreatment systems.

Fuel Cells

There is ample evidence of steady progress in most key fuel-cell-related technical areas, providing steady movement toward both performance and cost goals. There have been no breakthrough achievements, with the possible exception of a novel approach to the design and fabrication of the fuel cell membrane electrode assembly (MEA). The design, reported by 3M, eliminates the corrosion-prone carbon support structure and utilizes nanoscale metallic whiskers and a vacuum-deposited, thin film of catalyst. This approach, while not yet proven, offers the potential for simultaneously increasing fuel cell durability and reducing costs. The cost reductions would come from both a reduction in platinum loading and a configuration much more compatible with mass manufacturing. The performance increase would come primarily from better utilization of the catalyst.

Some other notable fuel cell achievements are these:

- The development of a reinforced membrane that improves durability with no apparent loss in performance;
- A better understanding of catalysts, especially platinum alloys, which

appear to have the potential for bringing as much as a tenfold improvement in activity; and

- The development of instrumentation and experimental procedures to allow real-time observation of water distribution in cells during transient operation.

However, there remain a number of barriers to viable fuel cell stacks, including these:

- Proven stack durability is only about one fourth of the targeted 5,000 hours.
- Cost, based on relatively proven technologies for the fuel cell system, is projected to be about four times the 2015 target of $30/kW. Note that the projected cost falls to about $67/kW, or about two times the target based on the novel but as yet unproven technology mentioned above.
- There are remaining performance barriers such as start time, especially at low temperatures.
- Predictable water management in the stack is critical and still difficult to achieve under all conditions.
- Virtually all hydrogen fuel cell vehicles are still operating on very high purity hydrogen. It is not yet clear what levels of contaminants can be tolerated without significant degradation of fuel cell performance or hardware lifetime.
- The membranes in the proton exchange membrane (PEM) systems are still limited to about 85°C, resulting in thermal management issues as well as some operational limitations.
- The impact of intake air quality on the life and durability of the electrocatalysts and fuel cell performance under on-road operating conditions are issues.
- There are newly recognized issues of catalyst chemical dissolution and stability and subsequent reprecipitation within the membrane.

Onboard Hydrogen Storage

This is another area where program achievements are notable but have not yet resulted in major progress toward storage system targets. The most significant of these achievements is the establishment of three centers of excellence (COEs), as well as the initiation some independent efforts. This has greatly improved the potential for isolating materials (if they exist) that might be suitable for onboard hydrogen storage systems.

The three hydrogen storage COEs are for (1) metal hydride, (2) hydrogen sorption, and (3) chemical hydrogen storage. The establishment of these three functional hydrogen storage COEs is an important achievement because each has reported substantial progress in the understanding of candidate materials. The organized and systematic approach of the COEs, with many researchers involved in common areas of investigation, clearly offers the best chance for success if, indeed, suitable materials exist.

Another notable achievement is the completion of extensive fast-fill tests for compressed gas storage to determine the circumstances under which precooling and/or communication between the hydrogen tank and the refueling system are needed. This is important since filling too fast can cause gas temperatures to exceed the safe limits for some tank materials (and components such as pressure relief devices), especially the resins that bind the carbon fibers and create structurally sound pressure vessels.

While considerable progress has been made, there are still very imposing barriers for achieving onboard hydrogen storage systems that will meet all targets and thus enable mass production of fuel-cell vehicles:

- To date, all demonstration fuel cell vehicles and, apparently, all planned next-generation fuel cell vehicles use either 350 bar (5,000 psi) or 700 bar (10,000 psi) compressed gas storage (except for a few vehicles using liquid hydrogen storage). There is wide agreement that compressed gas storage provides little opportunity for meeting either performance or cost targets, and liquid storage introduces many new problems in connection with the cryogenic temperature of –252°C (–423°F), including safety and boil-off issues. While with innovative vehicle and interface designs, compressed gas storage can provide a reasonable range and fill time, it does so at the expense of excessive volume, weight, and cost. For example,
  —Carbon fibers make up more than half of the weight and cost of compressed gas tanks, but little progress seems to have been made in reducing the cost of these fibers below \$25-\$35/kg.
  —Compressed gas tank temperatures are limited to about 85°C by the materials used. This necessitates precooling of the hydrogen and/or communication between the vehicle and the fueling station to fast-fill a nearly depleted 700-bar storage tank.

The investigation of solid or liquid storage materials as possible alternatives to compressed gas or liquid storage is also progressing, but with limited results to date. Specifically,

- While much progress is being made in understanding the potential of various materials for onboard hydrogen storage, no candidate materials have yet been identified that can meet system performance and cost targets.
- It has become clear from the studies that most of the materials that

appear to be capable of capturing and releasing sufficient quantities of hydrogen (hydrogen storage weight fraction) have either temperature or overall energy requirements incompatible with efficient operation of the storage system in conjunction with a PEM fuel cell.

Electrochemical Energy Storage

Many of the vehicle alternatives, especially plug-in hybrid electric vehicles (PHEVs), depend on affordable high-performance batteries. The most promising candidates seem to be lithium ion (Li ion) batteries:

- Li ion batteries can meet or exceed the weight, volume, power, and cycle life requirements for hybrid electric vehicles (HEVs).
- The development of abuse-tolerant electrodes, such as Li titanate anode material, which is also capable of high charge/discharge rates, is an important step for the success of these batteries.

While a great deal of progress has been made in Li ion battery technologies, there are still significant barriers:

- Li ion batteries still cost more than three times the $250/kW target.
- The durability of Li ion batteries has not been demonstrated, particularly the 15-year calendar life requirement.
- Current Li ion batteries are intolerant to abuse and could lead to safety issues. The development of abuse-tolerant electrodes such as the Li titanate anode mentioned above is promising in this regard but has not yet been demonstrated at full scale.

Safety, Codes and Standards

The development of national safety codes and standards is critical for the widespread operation of hydrogen-fueled vehicles. Every aspect of the operation of such vehicles, from onboard storage to refueling and even indoor parking, would be affected adversely by inadequate, inconsistent, or nonexistent codes and standards. Further, safety is of critical importance to maintaining support for the development of this technology. If hydrogen or hydrogen vehicles were ever demonstrated or perceived to be unsafe, this could be a severe blow to the FreedomCAR and Fuel Partnership.

Most of the achievements in the safety codes and standards are associated with the establishment of panels, databases, and handbooks that did not exist or had not been completed prior to 2005. Among the more notable are the following:

- The establishment of the DOE Hydrogen Technical Advisory Committee, whose activities are not, however, limited to safety, codes and standards.
- The publication of a hydrogen materials compatibility handbook (available online).
- Creation of a compendium of permitting tools.
- The formation of a hydrogen safety panel.
- The initiation of a hydrogen incidents database.
- The generation of hydrogen safety procedures for first responders.
- Experimentation and modeling of various hydrogen release and combustion scenarios.
- The publication online of a hydrogen bibliography.

Some of the potential barriers to achieving appropriate codes and standards are these:

- There is very little in the way of a hydrogen vehicle operational database to provide guidance.
- The multitude of authorities with jurisdiction complicates the setting of standards.
- Even in the best circumstances, developing codes and standards is a very slow process.

Vehicle Systems Analysis

The Phase 1 committee consistently recommended greater use of models, computer codes, and analyses. These tools provide guidance in screening materials and processes, planning test programs, and performing cost projections, as well as many other functions. Some of the tools that have been completed or updated are the following:

- *Greenhouse Gases, Regulated Emissions, and Energy Use in Transportation (GREET).* Tool for the analysis of vehicle configurations, capable of projecting source-to-wheels regulated emissions, energy consumption, and greenhouse gas emissions.
- *Powertrain Systems Analysis Toolkit (PSAT).* Used for evaluating vehicle technologies.
- *Hydrogen Technology Analysis-Hydrogen Production (H2A Production).* Model for projecting the production costs of hydrogen under various production scenarios.
- *Hydrogen Technology Analysis-Hydrogen Delivery (H2A Delivery).* Model for projecting the costs of delivering hydrogen using various delivery scenarios.

- *Hydrogen Transition (HyTrans).* Model for analyzing the transition to hydrogen-powered transportation. It includes issues relating to customer choice, vehicle market penetration, and governmental policy options.
- *National Energy Modeling System (NEMS).* General equilibrium model for projecting the effect of government policies associated with hydrogen production and utilization on the national economy.
- *Market Analysis (MARKAL).* Tool to project the impact of hydrogen production, supply infrastructure, and use of different feedstocks.
- *Hydrogen Logistics Model.* Tool to develop a strategy for minimizing the cost of delivered hydrogen by finding the most economical resources.

In addition to the tools described above, the Mobile Advanced Technology Testbed (MATT), a valuable tool for field evaluation of vehicles, has been completed and is in service.

Independent Cost Projections

In addition to cost projections associated with models such as H2A for the production and delivery of hydrogen, cost projections for the following have been completed or updated:

- *Vehicle fuel cell systems.* Projections were made by TIAX (an update) and Directed Technologies, Inc. (DTI) (new).
- *Compressed hydrogen onboard storage system.* Projections were made by TIAX.
- Distributed reforming of natural gas.

Hydrogen Production

Being able to produce and distribute hydrogen at a cost comparable to the costs of petroleum-based fuels is critical for the goals of the Partnership. Nearer term, production will probably rely on a combination of (1) distributed generation at forecourts employing electrolysis or the reforming of natural gas or bioderived fuels and (2) distribution from centralized sources. Longer term, centralized generation will grow because of lower costs and will most likely become the dominant source. So far, for long-term production only conversion of low-cost natural gas or coal has been reliably projected to cost less than $3.00/gge. DOE has shown that the United States could sustainably produce enough biomass to satisfy 30 percent or more of its current consumption of liquid transportation fuels if optimistic projections of biomass supply are met.

Some achievements in hydrogen production include these:

- Much better understanding of distributed generation of hydrogen and advanced sequestration through development of the FutureGen program.
- A better understanding of and ability to project of the amounts of biomass that could be made availabile for conversion to biofuels.
- Development of a redesigned electrolyzer with a projected reduction in cost, from $2,500/kW to $1,100/kW.
- Design of a high-pressure PEM electrolyzer capable of operating at 2,000 psi to eliminate a stage of compression.
- Concept for a low-cost alkaline electrolyzer with the potential to meet the 2012 capital cost target of $400/kW.
- Development of a delivery roadmap by the hydrogen delivery team.
- The completion of bench-scale testing of nuclear-based systems utilizing thermochemical or high-temperature electrolysis by the Office of Nuclear Energy (NE) with lab-scale testing expected to begin in September 2007.

Barriers to cost-competitive production include:

- Natural gas supply and price considerations are likely to restrict its use in the long term, as demand increases.
- The widespread use of coal depends on the availability of carbon sequestration, which has not yet been demonstrated.
- The projected capital cost of electrolyzers, while greatly reduced, is still about three times target values, and low-cost, nonpolluting electricity is not generally available for electrolyzers.
- Electrolyzers do not meet efficiency and durability targets.
- The sustainable availability and cost of biomass derived fuels are highly uncertain because of unresolved technical issues, unknowns surrounding land and water use policies, competition for these two resources, and the need for subsidies to stimulate commercial development.

Technology Validation

Experience teaches that the real-world operation of a system can result in unexpected consequences for its performance or durability. Thus it is very important to carefully monitor a technology to validate it. Two examples of such validation follow:

- *DOE vehicle/infrastructure demonstration.* Four teams representing 77 vehicles and 10 hydrogen stations are providing large quantities of real-world data on the operation and performance of the vehicles and the re-fueling operations. These data, which relate to the operation and performance of the vehicles as well as the refueling operations, are still being collected, but data collected so far have been presented by the National Renewable Energy Laboratory (NREL). Individual vehicles are not identified, but the composite ranges of data for critical variables are presented. The data are extremely important to researchers trying to move toward 2010 and 2015 targets.

- *Department of Transportation (DOT) fuel cell bus demonstrations.* Eight fuel cell buses are in operation and are providing data continuously.

Summary

There have been many important achievements in every area of the Partnership since the Phase 1 review, including some not mentioned here. Fuel cell technologies continue to advance, simultaneously reducing (projected) costs while improving performance. This provides the hope that such advances will continue until the targets are met. Advances are also evident in modeling, analysis, materials, and depth of understanding of the fundamental issues. Even so, there are many barriers remaining—including some that are not only very formidable but also potential roadblocks to the objectives of the program.

In the past, most program concerns centered on the fuel cell—indeed it is still very problematic. However, other barriers, such as finding an appropriate onboard hydrogen storage system, may have become more pressing. The reason is that while fuel cell technologies are continually advancing, a breakthrough of some kind seems to be needed to solve the storage problem.

It seems likely that the automotive original equipment manufacturers (OEMs) can innovate enough to store sufficient compressed hydrogen onboard for a 300-mile (or more) range, but it is not clear that this can ever be a satisfactory solution for millions of mass-produced vehicles. The hope in this area rests, to a great extent, on the combined talents and knowledge of the researchers at the newly established COEs to find acceptable storage materials and systems.

Other obvious areas of great concern are the production and dispensing of enough hydrogen to support large numbers of hydrogen-fueled vehicles.

Modeling and studies are beginning to identify the most important issues and provide direction, but here many of the potential roadblocks are already known and many more are sure to become known as the effort progresses. Reasonably accurate modeling is becoming so important that in almost every area there seems to be a need for an expanded knowledge base to allow additional analysis capabilities.

In summary, progress has been good in most areas and impressive in a few. However, resolving the barriers already known as well as those yet to be uncovered will clearly present major challenges.

ADEQUACY AND BALANCE OF THE PARTNERSHIP

DOE's total FY07 budget for hydrogen-related activities (the Hydrogen Fuel Initiative) is about $274 million, and total funding for activities relevant to the charter of the committee is about $401 million (Figure 5-1). The detailed allocation of these funds by main activity in the HFI is presented in Tables 5-1 and 5-2. Additional funding of about $98 million for FY07 was provided by industry and universities as part of the DOE-funded research, development, and demonstration activities. Other funding and resources from industry included about $16 million for Cooperative Research and Development Agreements (CRADAs) supporting Partnership goals. The private sector partners, of course, have significant proprietary programs with goals similar to those of the Partnership. The funding for these programs is not public knowledge, but in all it is reported to be at least twice the funding of the Partnership. The distribution of funding to performers—universities, private industry, national laboratories, and so on—is illustrated in Figures 5-2 and 5-3.

This level of expenditure is consistent with the priorities and recommendations of the NRC report *The Hydrogen Economy* and the DOE report *Hydrogen Posture Plan.* It is also consistent with the President's commitment of $1.7 billion over 5 years (FY04 to FY08) in his 2003 State of the Union message (NRC/NAE, 2004; DOE, 2004). The emphasis is on R&D related to fuel cell materials and

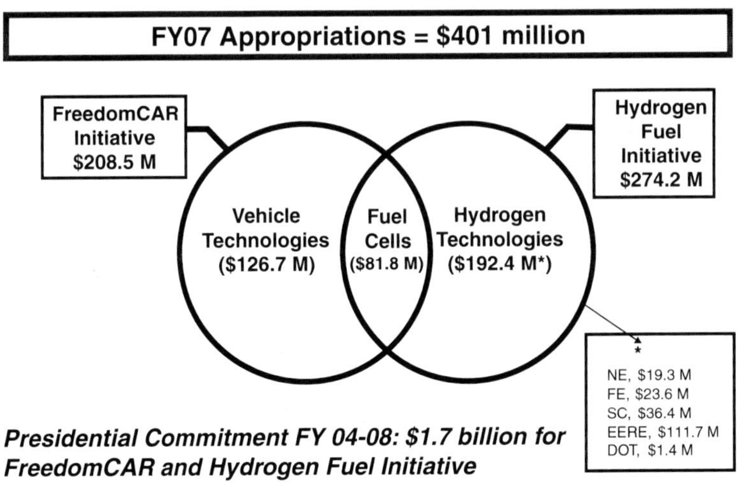

FIGURE 5-1 Estimated budget for the FreedomCAR and Fuel Partnership for FY07 Continuing Resolution. SOURCE: Phyllis Yoshida, DOE EERE, May 31, 2007.

TABLE 5-1 DOE Funding for Hydrogen Activities

	Funding (thousand $)			
			Continuing Resolution	
	Appropriated		Actual	Requested
Activity	FY05	FY06	FY07	FY08
Hydrogen production and delivery	13,303	8,391	34,594	40,000
Hydrogen storage R&D	22,418	26,040	34,620	43,900
Fuel cell stack component R&D	31,702	30,710	38,082	44,000
Technology validation	26,098	33,301	39,566	30,000
Transportation fuel cell systems	7,300	1,050	7,518	8,000
Distributed energy fuel cell systems	6,753	939	7,419	7,700
Fuel processor R&D	9,469	637	4,056	3,000
Safety codes and standards	5,801	4,595	13,848	16,000
Education	0	481	1,978	3,900
Systems analysis	3,157	4,787	9,892	11,500
Manufacturing R&D	0	0	1,978	5,000
Technical/program management support	535	0	0	0
Congressionally directed activities	40,236	42,520	0	0
Total	166,772	153,451	193,551	213,000

SOURCE: E. Wall and P. Davis, "Program overview," Presentation to the committee on April 25, 2007.

components, hydrogen production and delivery technology, and hydrogen storage materials. The budget also includes $50 million for basic science, which also agrees with the recommendations in *The Hydrogen Economy* that call for increased emphasis on the fundamental science related to hydrogen and fuel cell technologies. The budget also addresses the concern expressed in the NRC Phase 1 report by the Committee on Review of the FreedomCAR and Fuel Research Program, Phase 1 (NRC, 2005).

While hydrogen activity accounts, appropriately, for 70 percent of total program funding, there has been a significant increase in focus and additional assets allocated to nearer term opportunities such as HEVs, PHEVs, and advanced ICE combustion after a dip in such spending in FY06. The committee regards this change in balance as appropriate for three reasons: (1) It is in tune with the current

TABLE 5-2 Funding for the Hydrogen Fuel Initiative

	Funding (thousand $)			
			Continuing Resolution	
	Appropriated		Actual	Requested
Activity	FY05	FY06	FY07	FY08
DOE				
EERE hydrogen (HFCIT)	166,772	153,451	193,551	213,000
Fossil Energy (FE)	16,518	21,036	23,611	12,450
Nuclear Energy (NE)	8,682	24,057	18,665	22,600
Science (SC)	29,183	32,500	36,500	59,500
DOE subtotal	221,155	231,044	272,237	307,550
DOT	549	1,411	1,420	1,420
Total	221,704	232,455	273,747	308,975

SOURCE: E. Wall and P. Davis, DOE, "Program overview," Presentation to the committee on April 25, 2007.

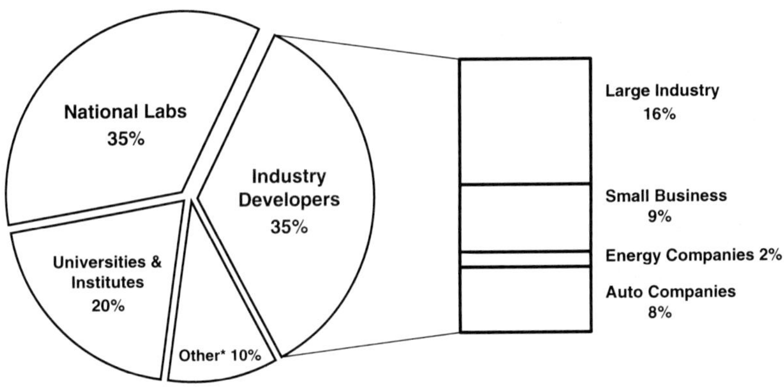

*Other includes SBIR/STTR and various crosscutting support activities, such as the Annual Merit Review and required EPAct05 studies and reports.

FIGURE 5-2 Distribution of $268 million total funding by recipient type for the DOE hydrogen program in FY07. SOURCE: Phyllis Yoshida, DOE EERE, November 19, 2007.

national dialogue on alternative energy; (2) it falls within the mission statement of the program; and (3) the resulting technologies will also be applicable to increasingly electrified vehicles and ultimately for fuel cell vehicles. (Much of the increased funding for these activities has come at the expense of the 21st Century Truck Partnership, which is beyond the scope of this committee).

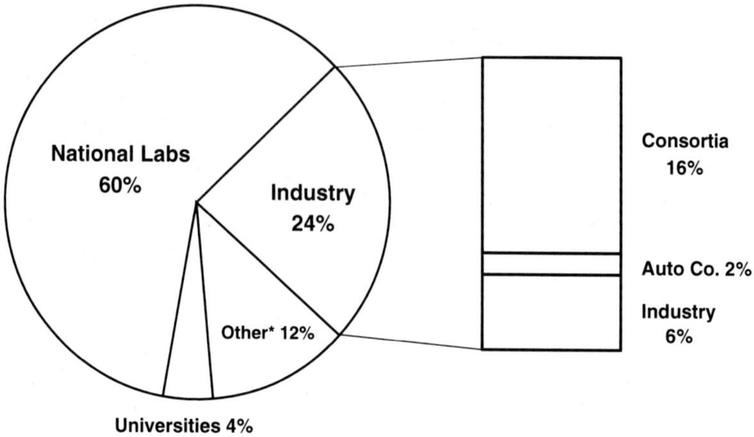

*Other includes SBIR/STTR, other government agencies, and various crosscutting support activities such as the Annual Merit Review and required EPAct05 studies and reports.

FIGURE 5-3 Distribution of $126.7 million total funding by recipient type for the vehicle technologies portfolio of the FreedomCAR and Fuel Partnership for FY07. SOURCE: Phyllis Yoshida, DOE EERE, November 19, 2007.

While the committee endorses the overall size and relative allocation strategy in the hydrogen program budget, there are five areas of concern. First, as discussed in Chapter 2, congressionally directed activities (earmarks) continue to negatively impact the program. The committee's Phase 1 report expressed concern at the number of earmarks in FY05, because they severely restricted the ability of DOE to effectively manage the program and delayed several of its important elements. Unfortunately, hydrogen program earmarks increased in FY06. Furthermore, for the first time, the Office of FreedomCAR and Vehicle Technologies (FCVT) budget was also affected by earmarks, which accounted for over 25 percent of the FY06 FCVT budget. It is serendipitous for the FreedomCAR and Fuel Partnership that FY07 has operated under a Continuing Resolution, in which there are no earmarks, and the committee would be grateful if this continued to be the case in FY08.

The second area of concern relates to the technology validation phase of the hydrogen program. The budget for this phase steadily increased through FY07 consistent with the deployment of increasing numbers of prototype fuel cell vehicles operating in diverse locations around the country. As described earlier in this chapter, this fleet of test vehicles is generating invaluable data on all aspects of hydrogen fuel cell vehicle operation, including the infrastructure. However, the technology validation budget request for FY08 has been reduced by 24 percent, and this will lead to a reduction in the number of vehicles deployed in extreme climates. The committee regrets this reduction, for two reasons: (1) It obviously constrains shared learning and (2) it is one area where the respective government and industry teams did not achieve consensus. The program is clearly most effective when it operates with the consensus of all the parties. The importance of maintaining a strong validation program cannot be overemphasized, and the committee urges DOE to reverse the proposed reduction in funding in FY08.

The third area of concern regarding the hydrogen program also carries over from the committee's Phase 1 report. While not directly within the purview of this committee, it is generally accepted that the feasibility of large-scale carbon capture and sequestration (CCS) essentially determines whether hydrogen can be produced from coal and/or natural gas in a future carbon-constrained environment and consequently affects the economics of hydrogen and its viability as a future fuel (energy carrier). Although DOE has sponsored a large number of pilot projects to explore CCS, the committee is concerned that the plans to monitor CO_2 leakage against the 99 percent retention goal are inadequate, especially as this is such a crucial aspect of CCS programs.

The fourth area of concern relates to safety, codes and standards: While the DOE activity in this area has increased significantly and is adequately funded, the DOT part of the program is well behind schedule and woefully underfunded. The National Highway Traffic Safety Administration (NHTSA) Four-Year Plan anticipated a budget of $4 million to $5 million per year, whereas current funding is only $1.4 million. It is recommended that DOT develop a long-range hydrogen safety plan with budget estimates and milestones to 2015 (see Chapter 2).

The fifth area of concern relates to the sustainable availability of biomass materials for conversion to hydrogen (and other fuels), as well as water and land requirements and the definition of subsidies that may be required. If the CCS program (noted above) is not completely successful, then biomass sources will become crucial, and this area deserves greater attention within the Partnership.

As noted earlier, focus and funding (shown in Table 5-3) within the vehicle technology portion of the program have been adjusted to emphasize hybrid vehicles, including PHEVs, and the committee endorses this emphasis. One kind of vehicle activity that the committee is inclined to challenge once again is the materials activity. After a 6 percent increase in FY07, the budget request for FY08 proposes to increase spending on structural materials another 12 percent, to almost $24 million, which is 19 percent of the total FreedomCAR vehicle expenditure. The work done to date by the materials team is excellent, but the committee continues to believe that the 50 percent weight reduction target at zero cost penalty is unrealistic and that funds currently allocated to this activity might be better spent elsewhere, as was suggested in the Phase 1 report.

In summary, there are five areas of concern for the Partnership, namely, congressionally directed activities (earmarks), the size of the technology validation

TABLE 5-3 DOE Funding for Vehicle Technologies Portion of the Freedom CAR and Fuel Partnership

	Funding (thousand $)		
Activity	FY06 Appropriations	FY07 Actual	FY08 Request
Hybrid electric systems	0	0	70,743
Vehicle systems	4,165	7,223	0
Hybrid and electric propulsion	41,023	64,841	0
Advanced combustion engine R&D	20,724	21,549	22,695
Materials technology	20,131	21,276	23,880
Fuels technology	7,041	10,085	7,001
Technology integration	0	0	2,300
Technology introduction	1,287	1,300	0
Innovative concepts	495	500	0
Technical/program management support	1,188	0	0
Biennial peer reviews	495	0	0
Congressionally directed activities	0	0	0
FreedomCAR and Fuel Partnership Total	96,549	126,774	126,619
21st Century Truck Partnership activities	45,267	45,020	29,792

SOURCE: Phyllis Yoshida, DOE EERE, June 8, 2007.

program, the design of the CCS pilot projects, the status of DOT safety, codes and standards activity, and the sustainable availability of biomass materials. The committee strongly supports the focus and allocation of funds within the vehicle portion of the program, with the exception of the spending on structural materials, which might be better used for some higher priority research areas.

Finally, the Partnership involves both short-term goals related to hydrocarbon-fueled vehicles used during a transition period and much longer term goals aimed at a clean and sustainable transportation energy future. The committee considers the current split of the funding between long-term and shorter-term goals to be appropriate. Hydrogen-related activities consume approximately 70 percent of the funds. The remaining funds support the development of transition technologies, where cost is often the most significant barrier, together with certain key technologies such as low-temperature combustion and enhanced battery performance.

OVERALL RESPONSE TO PHASE 1 RECOMMENDATIONS

This assessment focuses on the recommendations presented in the Executive Summary of the Phase 1 report (NRC, 2005). (See Appendix D in this report for a list of recommendations from the Phase 1 report.) The responses of the FreedomCAR and Fuel Partnership to the specific recommendations that were contained in Chapters 2, 3, and 4 of the Phase 1 report (Major Crosscutting Issues, Vehicle Subsystems, and Hydrogen Production, Delivery, and Dispensing) are addressed in the corresponding chapters of this report.

Fuel Cells and Hydrogen Storage

The following references to recommendation numbers can be found in Appendix D. Recommendations 3-6 and 3-9 emphasized fundamental research on membrane R&D, new catalyst systems, electrode design, and hydrogen storage. In particular, the Phase 1 report noted the risk posed to the hydrogen fuel cell vehicle program by reliance on high pressure storage beyond the early transition period. Even with many automotive manufacturers currently introducing fuel cell vehicles that employ high-pressure tanks, the potential for low-pressure hydrogen storage to accelerate a hydrogen transition remains enormous. This was a major concern in the Phase 1 report, and it remains one in this report.

The committee recognizes the actions that the Partnership has taken to address these fuel cell and hydrogen storage issues. It notes that they are ongoing priorities and that their successful resolution will require that this effort extend throughout the hydrogen transition.

Electrochemical Energy Storage for Electric Vehicles

Recommendation 3-11 proposed that high-energy batteries be given higher priority. The Partnership concurs, and funding for breakthrough research has increased markedly. The analyses of this committee continue to confirm the importance of battery technology, which is essential for success of battery electric vehicle (EVs), HEVs, PHEVs, and hydrogen fuel cell vehicles. Consider, for example, the joint announcement on July 10, 2007, of Ford and Southern California Edison for a multiyear PHEV evaluation and demonstration program. Toyota has also announced a PHEV collaboration with the University of California. These programs will elicit much information about the performance of these vehicles in the hands of consumers and about their interaction with the stationary electric system; however, the commercial market must await lower-cost, high-energy batteries.

Electrical Systems and Electronics

All-electric-drive vehicles must successfully integrate the systems that manage the flow of electric energy from its multiple possible sources (off-board connections to the electric grid, onboard generator, regenerative braking, and so forth) to its multiple uses (torque at the wheels, passenger comfort, battery charging, information, and so forth). Recommendations 3-16, 3-17, and 3-18 proposed that the electrical and electronic systems technical team coordinate the diverse research activities pertaining to electrical systems with the aim of achieving significant cost advantages. The Partnership has concurred and has begun that process in coordination with the DOE systems analysis activity. The committee continues to support this electronic systems integration as a vital strategic goal.

Hydrogen Fuel Production and Distribution

Under Recommendation 4-2, the committee called for special attention to be directed at the transition from a fuels infrastructure built to serve ICEs to one capable of serving a mixed fleet. In particular, the systems analysis work supporting the fuel/vehicle pathway integration technical team should examine whether raising the cost goals for hydrogen production during the transition period would accelerate or retard the introduction of hydrogen fuel cell vehicles. These analyses have begun but have not yet been completed. The committee continues to urge attention to this vital component of a hydrogen fuel cell vehicle roll out strategy.

In Recommendation 4-3, the committee proposed greater attention to distributed hydrogen production, including by both natural gas reforming and electrolysis, as well as exploratory work on other distributed production options. As of this writing, DOE has focused on electrolysis and reforming. The committee continues to suggest exploratory research into hydrogen production at the forecourt that would use feedstocks other than water and natural gas and that might compete successfully in a mature hydrogen economy.

In Recommendation 4-5, the committee suggested creating a CCS subteam. In response, the Partnership pointed out that the hydrogen production technical team has this responsibility and coordinates closely with DOE's Office of Fossil Energy, which manages the CCS program for DOE. Noting the importance of this liaison, the committee believes this arrangement can be made to work satisfactorily with ongoing management attention. However, it remains concerned that the CCS program will not deliver results rapidly enough to meet the key decision points in the hydrogen program.

Structural Materials

Recommendation 3-21 noted that more extensive research on carbon-fiber-reinforced polymers and direct cooperation with the principal fiber manufacturers will be essential for meeting the FreedomCAR and Fuel Partnership goals. R&D

on manufacturing vehicle structures should continue. The committee appreciates the Partnership's emphasis on this research because of its importance as a hedge against delays in the commercial introduction of low-pressure, on-vehicle hydrogen storage.

Recommendation 3-25 proposed a review of DOE expenditures on materials research to see if the resources could be used in higher priority research elsewhere—fuel cells, hydrogen storage materials or batteries, for example. DOE conducted such a review and concluded that support for lightweight materials should not be redirected elsewhere. In view of this program decision, the committee now recommends a review of the cost goals for lightweight materials with the intent of gaining a more realistic understanding of what can be achieved. The committee also continues to recommend that these funds should for the most part be redirected to higher priority research elsewhere (see Chapter 3) except for projects that show great promise for enabling, near term and at low cost, a reduction in mass.

Crosscutting Issues

Safety

Recommendation 2-5 recommended that the Partnership form a new crosscutting technical team to address broad hydrogen-related safety issues. The committee further recommended increasing resources not only from the FreedomCAR and Fuel Partnership but also from the other participating federal agencies, chiefly NHTSA. DOE requested the needed funds, but its subsequent review of this recommendation concluded that a separate technical team could not function as envisioned by the committee and declined to establish a new technical team. While the committee must defer to DOE in matters of government organization, several observations should nevertheless remain before the management of the FreedomCAR and Fuel Partnership:

- Safety is best addressed before costly recalls must be made and the Partnership's reputation has been damaged.
- The Learning Demonstration Program can become an effective tool for identifying incipient safety issues.
- Where the statutory responsibility requires other branches of the federal government to become involved in hydrogen safety, DOE should exercise leadership to ensure that these efforts are adequately supported. DOT needs to prepare a long-range hydrogen safety plan and work to get it adequately funded.

Public Concerns

Recommendation 2-16 recommended that DOE collaborate with the Environmental Protection Agency to systematically identify and examine the consequences of widespread hydrogen production and use. DOE concurred and is using the Programmatic Environmental Impact Statement process as the backbone for this assessment. The committee recognizes the scope and breadth of the DOE response. As with safety, environmental impacts are better recognized and addressed early in the program rather than discovered after large-scale investments have been made.

Systems Analysis

Recommendation 2-2 proposed that the Partnership should use its systems analysis capabilities routinely in all management activities—establishing goals, evaluating trade-offs, setting priorities, and making go/no-go decisions. Recommendation 2-1 emphasized a specific element of this, the use of ongoing well-to-wheels analyses to assess progress in the FreedomCAR and Fuel Partnership and to guide trade-offs among goals.

DOE has concurred, and the committee recognizes the progress that has been made since the 2005 report. The committee continues to encourage the further integration of the systems approach into all aspects of program management, both as a guide to effective management and as a way to communicate with the diverse set of the Partnership stakeholders.

Strategy for Accomplishing Goals

The Phase 1 committee also recommended, following on Recommendation 2-2, that the Partnership should perform an overall program evaluation using go/no-go decisions and setting priorities focused on the most important goals. DOE has concurred. Looking ahead, the committee recognizes that the future of the FreedomCAR and Fuel Partnership beyond 2008 remains to be determined. Nevertheless, the committee recommends that the Executive Steering Group begin a strategic planning activity that would establish the most important objectives and ensure the means to achieve them.

REFERENCES

Department of Energy (DOE). 2004. *Hydrogen Posture Plan: An Integrated Research, Development and Demonstration Plan*. Washington, D.C.: U.S. Department of Energy. Available on the Web at <http://www.eere.energy.gov/hydrogenandfuelcells/pdfs/hydrogen_posture_plan.pdf>.

National Research Council (NRC). 2005. *Review of the Research Program of the FreedomCAR and Fuel Partnership, First Report*. Washington, D.C.: The National Academies Press.

National Research Council/National Academy of Engineering (NRC/NAE). 2004. *The Hydrogen Economy: Opportunities, Costs, Barriers, and R&D Needs*. Washington, D.C.: The National Academies Press.

Appendixes

A

Organization Chart for the U.S. Department of Energy's Office of Energy Efficiency and Renewable Energy (as of July 10, 2007)

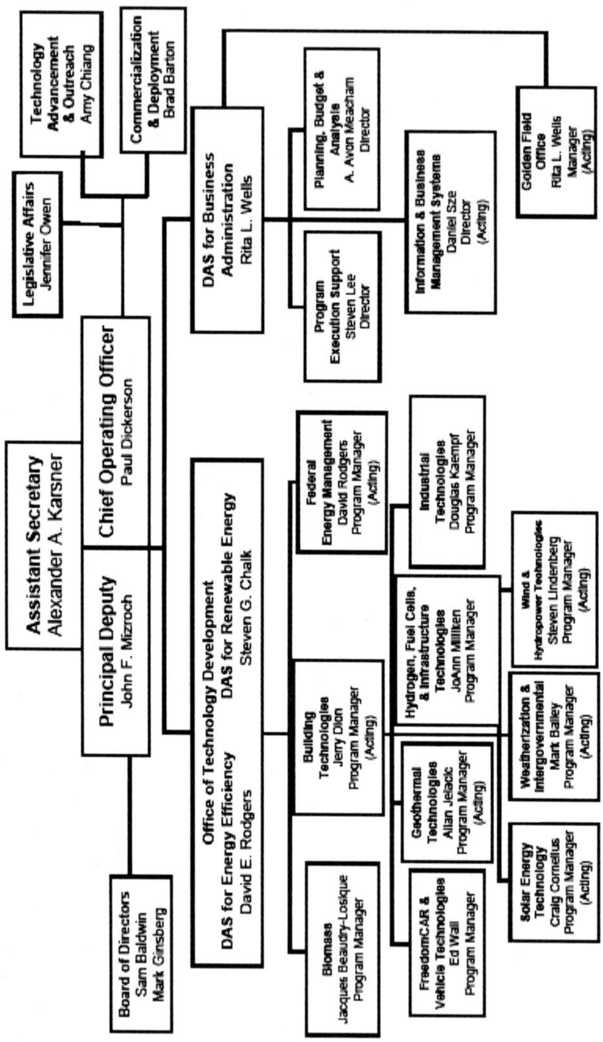
Assistant Secretary
Alexander A. Karsner
Principal Deputy
John F. Mizroch
Chief Operating Officer
Paul Dickerson
Board of Directors
Sam Baldwin
Mark Ginsberg
Legislative Affairs
Jennifer Owen
Technology Advancement & Outreach
Amy Chiang
Commercialization & Deployment
Brad Barton
Office of Technology Development
DAS for Energy Efficiency
David E. Rodgers
DAS for Renewable Energy
Steven G. Chalk
DAS for Business Administration
Rita L. Wells
Program Execution Support
Steven Lee
Director
Planning, Budget & Analysis
A. Avon Meacham
Director
Information & Business Management Systems
Daniel Sze
Director
(Acting)
Golden Field Office
Rita L. Wells
Manager
(Acting)
Biomass
Jacques Beaudry-Losique
Program Manager
Building Technologies
Jerry Dion
Program Manager
(Acting)
Federal Energy Management
David Rodgers
Program Manager
(Acting)
FreedomCAR & Vehicle Technologies
Ed Wall
Program Manager
Geothermal Technologies
Allan Jelacic
Program Manager
(Acting)
Hydrogen, Fuel Cells, & Infrastructure Technologies
JoAnn Milliken
Program Manager
Industrial Technologies
Douglas Kaempf
Program Manager
Solar Energy Technology
Craig Cornelius
Program Manager
(Acting)
Weatherization & Intergovernmental
Mark Bailey
Program Manager
(Acting)
Wind & Hydropower Technologies
Steven Lindenberg
Program Manager
(Acting)

B

Biographical Sketches of Committee Members

Craig Marks (NAE), *Chair,* is a member of the Board of Trustees of Altarum, a not-for-profit research firm focused on shifting health-care spending in the United States toward systems that are centered on the value of health. For 27 years he worked at General Motors in engineering and management positions. He subsequently became vice president of Engineering and Technology for the TRW Automotive Sector and then Vice President of Technology and Productivity for the Allied Signal Automotive Sector. In the latter position he headed an automotive R&D center and was responsible for the staff functions of manufacturing, quality, health, safety and environment, and communications. After retiring, Dr. Marks became an adjunct professor at the University of Michigan, with a joint appointment in the College of Engineering and the School of Business Administration, where he helped found the Joel D. Tauber Manufacturing Institute. He has served on a number of NRC committees including as chairman, Committee on Review of the Research Program of the Partnership for a New Generation of Vehicles, and chairman, Committee for the Review of the Intelligent Vehicle Initiative—Phase 2. He is a member of the National Academy of Engineering and a fellow of the Society of Automotive Engineers (SAE) and the Engineering Society of Detroit. Dr. Marks holds B.S., M.S., and Ph.D. degrees in mechanical engineering from the California Institute of Technology in Pasadena.

Peter Beardmore (NAE) was formerly director, Ford Research Laboratory, Ford Motor Company, prior to his retirement in August 2000. His primary research

Note: NAE = member, National Academy of Engineering.

interests are the deformation and fracture of materials, including extensive research experience in metals, polymers, and composites, and he has published over 83 technical articles. He is a recognized international authority on composite materials and on the application of new materials to automotive structures. His management responsibilities at Ford covered a wide area of research activities relative to the automotive industry, including materials, environmental chemistry, sensor technologies, automotive catalyst development, and the application of modern analytical techniques. He is a member of the American Society for Materials (ASM), The Metallurgical Society (TMS) of AIME, and the Engineering Society of Detroit (ESD). He was elected a fellow of ASM in 1989 and a fellow of ESD in 1991. In 1992, he was elected a member of the National Academy of Engineering. He holds a B.Met. in metallurgy from the University of Sheffield and a Ph.D. in metallurgy from the University of Liverpool.

David L. Bodde serves as a professor and senior fellow at Clemson University. There, he directs innovation and strategy at Clemson's International Center for Automotive Research. Prior to joining Clemson University, Dr. Bodde held the Charles N. Kimball Chair in Technology and Innovation at the University of Missouri in Kansas City. Dr. Bodde serves on the board of directors of several energy and technology companies, including Great Plains Energy and the Commerce Funds. His executive experience includes vice president, Midwest Research Institute; assistant director of the Congressional Budget Office; and deputy assistant secretary in the U.S. Department of Energy. Dr. Bodde frequently testifies before congressional committees. He was once a soldier and served in the Army in Vietnam. He has a doctorate in business administration from Harvard University, M.S. degrees in nuclear engineering (1972) and management (1973), and a B.S. from the United States Military Academy.

Glenn A. Eisman is a research professor in materials science and engineering at Rensselaer Polytechnic Institute in Troy, N.Y. He is also adjunct professor at the graduate school of Union University (Schenectady, N.Y.); principal partner of Eisman Technology Consultants, LLC; and managing partner of H2 Pump LLC (Niskayuna, N.Y.). His previous positions include chief technology officer, Plug Power, Inc.; technical leader, The Advanced Materials Program, Central Research and New Businesses, The Dow Chemical Company; project leader, Discovery Research R&D and Inorganic Chemicals Research, The Dow Chemical Company; and Robert A. Welch Research Fellow, The University of Texas-Austin. Dr. Eisman has extensive experience in R&D and product development on fuel cells, hydrogen technologies, electrochemical engineering, physical and inorganic solid state chemistry, and new technology commercialization and business development. He received the Inventor of the Year Award from Dow Chemical (1993) and is a member of the Electrochemical Society. He received his bachelor's degree in chemistry from Temple University and a Ph.D. in physical inorganic chemistry from Northeastern University.

W. Robert Epperly is a consultant. From 1994 to 1997, he was president of Catalytica Advanced Technologies, Inc., a company developing new catalytic technologies for the petroleum and chemical industries. Prior to joining Catalytica, he was CEO of Fuel Tech N.V., a company specializing in new products for combustion and air pollution control. Earlier, he was general manager of Exxon Corporate Research. While at Exxon Research and Engineering Co., he was also general manager of the Synthetic Fuels Department, where he was responsible for the engineering of commercial projects, and manager of the Baytown Research and Development Division, where he was responsible for coal conversion research. Earlier, he was director of the Fuels Research Laboratory, responsible for R&D on Exxon's fuels products. He is a fellow in the American Institute of Chemical Engineers and a past recipient of the AIChE's National Award in Chemical Engineering Practice. He has authored or coauthored over 50 publications, including two books, and has 38 U.S. patents. He has broad experience in the conversion of fossil resources to alternative gaseous and liquid fuels, petroleum fuels, catalysis, air pollution control, and R&D management. Since 1981, he has participated on nine committees at the National Research Council, including the Committee on Alternatives and Strategies for Future Hydrogen Production and Use. He received B.S. and M.S. degrees in chemical engineering from Virginia Tech.

David E. Foster is the Phil and Jean Myers Professor of Mechanical Engineering, University of Wisconsin, Madison, and former director of the Engine Research Center, which has won two center of excellence competitions for engine research and has extensive facilities for research on internal combustion engines. A member of the faculty at the University of Wisconsin since he completed his Ph.D., Dr. Foster teaches and conducts research in thermodynamics, fluid mechanics, internal combustion engines, and emission formation processes. His work has focused specifically on perfecting the application of optical diagnostics in engine systems and the incorporation of simplified or phenomenological models of emission formation processes into engineering simulations. He has published more than 70 technical articles in this field throughout the world and for leading societies in this country. He is a recipient of the Ralph R. Teetor Award, the Forest R. McFarland Award, and the Lloyd L. Withrow Distinguished Speaker Award of the Society of Automotive Engineers (SAE), and he is an SAE fellow and has been awarded the ASME Honda Gold Medal. He has served on a number of NRC committees, including the Committee on Review of the Research Program of the Partnership for a New Generation of Vehicles. He is a registered professional engineer in the State of Wisconsin and has won departmental, engineering society, and university awards for his classroom teaching. He received a B.S. and an M.S. in mechanical engineering from the University of Wisconsin and a Ph.D. in mechanical engineering from the Massachusetts Institute of Technology (MIT).

John B. Heywood (NAE) is Sun Jae Professor of Mechanical Engineering at MIT and director of the Sloan Automotive Laboratory. Dr. Heywood's research has focused on understanding and explaining the processes that govern the operation and design of internal combustion engines and their fuels requirements. Major research activities include engine combustion, pollutant formation, operating and emissions characteristics and fuel requirements of automotive engines, and assessing future propulsion system developments. He has served on a number of NRC committees, including the Committee on Review of the Research Program of the Partnership for a New Generation of Vehicles. He has consulted for many companies in the automotive and petroleum industries and for government organizations. He has received many awards, from the American Society of Mechanical Engineers, the British Institution of Mechanical Engineers, and the Society of Automotive Engineers for his research contributions. He has a Ph.D. in mechanical engineering from MIT, a Sc.D. from Cambridge University, and honorary doctorates from Chalmers University of Technology (Sweden) and City University (U.K.).

Harold H. Kung is professor of chemical and biological engineering at Northwestern University. His areas of research include surface chemistry, catalysis, and chemical reaction engineering. His professional experience includes work as a research chemist at E.I. du Pont de Nemours & Co., Inc. He is a recipient of the P.H. Emmett Award and the Robert Burwell Lectureship Award from the North American Catalysis Society, the Herman Pines Award of the Chicago Catalysis Club, the Cross-Canada Lectureship from the Catalysis Division of the Chemical Institute of Canada, and is a Catalysis Society of South Africa eminent visitor. He is a fellow of the American Association for the Advancement of Science, and editor of *Applied Catalysis A: General*. He has a Ph.D. in chemistry from Northwestern University.

James J. MacKenzie is a senior fellow in the World Resources Institute's (WRI's) Climate, Energy, and Pollution program. Prior to joining WRI, Dr. MacKenzie was a senior staff scientist, Union of Concerned Scientists; a senior staff member for energy, President's Council on Environmental Quality (CEQ); and a member of the joint scientific staff of the Massachusetts and national Audubon Societies. Much of his recent research and analysis has focused on transportation technologies and the impact of the transportation system on the environment. He is co-author (transportation chapter) of *Frontiers of Sustainability: Environmentally Sound Agriculture, Forestry, Transportation, and Power Production*; author of *Climate Protection and the National Interest*; *Oil as a Finite Resource: When Is Global Production Likely to Peak?*; and *The Keys to the Car, Electric and Hydro-*

gen Vehicles for the 21st Century. He is also co-author of *Car Trouble*, a book on the impacts of cars on the American scene, and of several major WRI reports, including an analysis of the subsidies for motor vehicles in the United States, the impacts of global motor vehicle use on climate change, and the effects of multiple air pollutants on U.S. forests and crops. He has also completed a policy report exploring the linkages among the problems of climate change, air pollution, and national energy security. Dr. MacKenzie received his Ph.D. in physics from the University of Minnesota and completed postgraduate work at Los Alamos and Argonne National Laboratories and MIT before joining the Audubon Society.

Christopher L. Magee (NAE) is professor, Engineering Systems Division, Massachusetts Institute of Technology, and co-director, Engineering Design and Advanced Manufacturing for the MIT-Portugal Program. Prior to joining MIT, he held a number of positions at Ford Motor Company, including director, Vehicle Systems Engineering; director, Advanced Vehicle Engineering; manager, Materials Science Department; senior research scientist, Metallurgy Department; and executive director, Programs and Advanced Engineering, with global responsibility for all major technically deep areas involved in Ford's Product Development Organization. He has expertise in such areas as phase transformations, plastic deformation, materials strength, large-scale collapse of engineering structures, product development, automotive design, value engineering, and simultaneous manufacturing/product engineering. He has made important contributions to the understanding of the transformation, structure, and strength of ferrous materials and to lightweight materials development and implementation; he pioneered experimental work on high-rate structural collapse aimed at vehicle crashworthiness; and he adapted systems engineering to the modern automotive design process. His recent research has emphasized innovation and technology development in complex systems. He was elected to the National Academy of Engineering for contributions to advanced vehicle development, was a Ford Technical Fellow (1996), and is a fellow of ASM. He has a B.S., an M.S., and a Ph.D. in metallurgy and materials science from the Carnegie Institute of Technology (now Carnegie Mellon University) and an M.B.A. from Michigan State University.

Robert J. Nowak is a private consultant. He was a program manager at the Defense Advanced Research Projects Agency and the Office of Naval Research. He has directed and supported research in fundamental electrochemistry, fuel cells, batteries, capacitors, energy harvesting, fuel processing, thermal energy conversion, micro-engines, hydrogen storage, biofuel cells, sonoluminescence, and biomolecular motors. He recently served on an EPRI committee to evaluate direct carbon fuel cell technologies, the NRC Committee on Portable Energy Sources for the Objective Force Warrior, and the NRC Panel on Benefits of Fuel Cell R&D. He received his B.A. and M.S. degrees in chemistry from Oakland University and his Ph.D. degree in chemistry from the University of Cincinnati. He was a postdoctoral research associate with Professor Royce W. Murray at the University of North Carolina at Chapel Hill and he was selected as NRC postdoctoral fellow at the Naval Research Laboratory, where he continued his research activities in conducting polymer electrochemistry and chemically modified electrodes as a staff scientist and section head. Dr. Nowak received the Secretary of Defense Meritorious Civilian Service Award in 2002 for his efforts in developing portable power sources for the military.

Michael P. Ramage (NAE) is retired executive vice president, ExxonMobil Research and Engineering Company. Previously he was executive vice president and chief technology officer, Mobil Oil Corporation. Dr. Ramage held a number of positions at Mobil, including research associate, manager of process research and development, general manager of exploration and producing research and technical service, vice president of engineering, and president of Mobil Technology Company. He has broad experience in many aspects of the petroleum and chemical industries. He has served on a number of university visiting committees and was a member of the Government-University Industrial Research Roundtable. He was a director of the American Institute of Chemical Engineers and is a member of several professional organizations. Dr. Ramage chaired the recent NRC report *The Hydrogen Economy: Opportunities, Costs, Barriers, and Research Needs*. He is a member of the National Academy of Engineering and has served on the NAE Council. Dr. Ramage has B.S., M.S., Ph.D., and HDR degrees in chemical engineering from Purdue University.

Vernon P. Roan is retired director of the Center for Advanced Studies in Engineering and professor of mechanical engineering at the University of Florida, where he has been a faculty member for more than 30 years. Since 1994, he has also been the director of the University of Florida Fuel Cell Research and Training Laboratory. Previously, he was a senior design engineer with Pratt and Whitney Aircraft. Dr. Roan, who has more than 25 years of research and development experience, is currently working as a consultant to Pratt & Whitney on advanced gas-turbine propulsion systems. His research at the University of Florida has involved both spark-ignition and diesel engines operating with many alternative fuels and advanced concepts. With groups of engineering students, he designed and built a 20-passenger diesel-electric bus for the Florida Department of Transportation and a hybrid-electric urban car using an internal combustion engine and lead-acid batteries. He has been a consultant to the Jet Propulsion Laboratory, monitoring its electric and hybrid vehicle programs. He has organized and chaired two national meetings on advanced vehicle technologies and a national seminar on the development of fuel-cell-powered automobiles and has published numerous technical papers on innovative propulsion systems. He was one of the four members of the Fuel Cell Technical Advisory Panel of the California Air Resources Board (CARB), which issued a report in May 1998 on the status and outlook for fuel cells for transportation applications. He is currently a member of the Expert Panel on Zero Emission Vehicles for CARB. Dr. Roan received a B.S. in aeronautical engineering, an M.S. in engineering from the University of Florida, and a Ph.D. in engineering from the University of Illinois.

Bernard Robertson (NAE) is president of BIR1, LLC, an engineering consultancy specializing in transportation and energy matters that he founded in January 2004, upon his retirement from DaimlerChrysler Corporation. During the latter part of his 38-year career in the automotive industry, Mr. Robertson was elected an officer of Chrysler Corporation in February 1992. He was appointed senior vice president when Chrysler Corporation and Daimler-Benz AG merged in November 1998, and was named senior vice president of engineering technologies and regulatory affairs in January 2001. In his last position, he led the Liberty and Technical Affairs Research group, Advanced Technology Management and FreedomCAR activities, and hybrid electric, battery electric, fuel cell, and military vehicle development. In addition, he was responsible for regulatory analysis and compliance for safety and emissions. Mr. Robertson holds an M.B.A. degree from Michigan State University, a master's degree in automotive engineering from the Chrysler Institute, and a master's degree in mechanical sciences from Cambridge University, England. He is a member of the National Academy of Engineering, a fellow of the Institute of Mechanical Engineers (U.K.), a chartered engineer (U.K.), and a fellow of the Society of Automotive Engineers.

R. Rhoads Stephenson is currently a technology consultant. Previously, he held a number of positions at the Jet Propulsion Laboratory (JPL), the National Highway Traffic Safety Administration (NHTSA), and Martin Marietta Corporation. At JPL, these included deputy director and acting director, Technology and Applications Programs; manager, Electronics and Control Division; deputy manager, Control and Energy Conversion Division; and manager of the Systems Analysis Section. He also served as associate administrator for R&D at NHTSA, and while at Martin Marietta Corporation worked on energy conversion devices for space power. He has been a consultant to the Motor Vehicle Fire Research Institute, has been providing peer reviews of automotive safety issues, and has recently published a number of papers on crash-induced fire safety issues with motor vehicles, including hydrogen-fueled vehicles. He brings extensive expertise in vehicle safety analysis, advanced technology systems, energy conversion technologies, and energy and environmental analysis. He has B.S., M.S., and Ph.D. degrees in mechanical engineering from Carnegie Mellon University.

Kathleen C. Taylor (NAE) is retired director of the Materials and Processes Laboratory at General Motors Research and Development and Planning Center in Warren, Michigan. Dr. Taylor was simultaneously chief scientist for General Motors of Canada, Ltd., in Oshawa, Ontario. Earlier Dr. Taylor was department head for physics and physical chemistry and department head for environmental sciences. She serves at the Catalysis Society, the Board of Directors of the National Inventors Hall of Fame, the DOE Basic Energy Sciences Advisory Committee, and was formerly a member of the NRC Board of Energy and Environmental Systems. Dr. Taylor was awarded the Garvan Medal from the American Chemical Society. She is a member of the National Academy of Engineering and a foreign fellow of the Indian National Academy of Engineering. She is a fellow of SAE International and the American Association for the Advancement of Science. She has been president of the Materials Research Society and chair of the board of directors of the Gordon Research Conferences. She has expertise in R&D management, fuel cells, batteries, catalysis, exhaust emissions control, and automotive materials. She received an A.B. in chemistry from Douglass College and a Ph.D. in physical chemistry from Northwestern University.

Giri Venkataramanan is associate professor, Department of Electrical and Computer Engineering, University of Wisconsin, Madison, and associate director, Wisconsin Electric Machines and Power Electronics Consortium. Previous positions include associate professor, Montana State University, Bozeman; visiting research associate, Lawrence Berkeley National Laboratory; and visiting researcher, CNPq, Brazilian National Council for Development of Science and Technology, Federal University of Minas Gerais, Brazil. His fields of interest include electrical power conversion, AC power flow control, design of power converters, distributed generation, power converter architecture, and power converter packaging. Specific research projects focus on characterization of power semiconductor devices and components, development of novel power converters and control strategies, physical realization and packaging, mitigation of converter-induced harmonics, and control of electromagnetic interference. He is active in a number of IEEE technical forums, and served as chair, IEEE Montana Section. He participated in an NRC workshop held by the Committee on Assessment of Combat Hybrid Power Systems. He has a B.E. from the University of Madras, India, in electrical and electronics engineering, an M.S. from the California Institute of Technology, and a Ph.D. from the University of Wisconsin, Madison, in electrical engineering.

Brijesh Vyas is currently distinguished member of technical staff in the Nanofabrication Research Department at Bell Laboratories–Lucent Technologies. Earlier he was the technical manager of the Energy Conversion Technology Group at Bell Laboratories. He also held positions at Brookhaven National Laboratory and the Technical University of Denmark. His primary responsibility is the application of electrochemical technologies to nanofabrication. He has been responsible for R&D of advanced materials and technologies for high-energy batteries for portable applications and forward-looking work on energy storage systems for standby applications including batteries, fuel cells, flywheels, and photovoltaic devices. He has led efforts on R&D for capacitors and for rechargeable lithium, nickel cadmium, nickel metal hydride, and lead acid batteries. In addition he has been responsible for battery technology transfer to manufacturing and interacted with application engineers and marketing and legal organizations. He is a recipient of the Sam Tour award by the American Society of Testing and Materials and is a member of the Electrochemical Society. He served on the NRC Committee to Review the U.S. Advanced Battery Consortium's electric vehicle battery R&D project selection process. He holds a B.Tech. in metallurgical engineering from the Indian Institute of Technology–Bombay and a Ph. D. in materials science from the State University of New York, Stony Brook.

C

Presentations and Committee Meetings

COMMITTEE MEETING, SOUTHFIELD, MICHIGAN, MARCH 1, 2007

Opening Remarks
Larry Burns, General Motors

Perspective on the FreedomCAR and Fuel Partnership
Steve Zimmer, DaimlerChrysler
Maria Curry-Nkansah, BP

Overview of the FreedomCAR and Fuel Partnership
Ed Wall and JoAnn Milliken, U.S. Department of Energy

Vehicle Systems Analysis
Asi Perach, Ford
Lee Slezak, U.S. Department of Energy

Fuel Pathway Integration
Karel Kapoun, Shell
Fred Joseck, U.S. Department of Energy

Advanced Combustion and Emissions Control
Richard Peterson, General Motors
Ken Howden, U.S. Department of Energy

Electrochemical Energy Storage
Ahsan Habib, General Motors
Dave Howell, U.S. Department of Energy

Fuel Cells
Doanh Tran, DaimlerChrysler
Kathi Epping, U.S. Department of Energy

Electrical and Electronics
Vijay Garg, Ford
Susan Rogers, U.S. Department of Energy

Materials
Andy Sherman, DaimlerChrysler
Joe Carpenter, U.S. Department of Energy

Onboard Hydrogen Storage
Scott Jorgensen, General Motors
Sunita Satyapal, U.S. Department of Energy

Hydrogen Production
Steve Schlasner, ConocoPhillips
Roxanne Garland, U.S. Department of Energy

Hydrogen Delivery
Jim Kegerreis, ExxonMobil
Mark Paster, U.S. Department of Energy

Codes and Standards
Jesse Schneider, DaimlerChrysler
Pat Davis, U.S. Department of Energy

COMMITTEE MEETING, WASHINGTON, D.C., APRIL 25-26, 2007

Program Overview
Ed Wall and Pat Davis, U.S. Department of Energy

Systems Analysis Efforts
Fred Joseck and Lee Slezak, U.S. Department of Energy

Well-to-Wheels Analysis
Michael Wang, Argonne National Laboratory

Learning Demos/NREL Data
Keith Wipke, National Renewable Energy Laboratory
Sig Gronich, U.S. Department of Energy

Wind/Biomass Production of Hydrogen
National Renewable Energy Laboratory

Nuclear Production of Hydrogen
Carl Sink, U.S. Department of Energy

Fossil Energy Production of Hydrogen
Lowell Miller, U.S. Department of Energy

Basic Energy Science (BES)
Harriet Kung, U.S. Department of Energy

Safety, Codes and Standards
Jay Keller, Sandia National Laboratories
Pat Davis, U.S. Department of Energy

TIAX Cost Analysis
Steve Lasher, TIAX LLC

COMMITTEE MEETING, WASHINGTON, D.C., JUNE 27-28, 2007

PSAT: Status and Use for Managing the Program Tradeoffs
Lee Slezak, U.S. Department of Energy
Aymeric Rousseau, Argonne National Laboratory

Energy Storage Materials—Results of BES Workshop
Harriet Kung, U.S. Department of Energy

SBIR Activities
Ed Wall and JoAnn Milliken, U.S. Department of Energy

DOT Hydrogen Safety Activities
William Chernikoff, U.S. Department of Transportation

DOE Office of Biomass/Biofuels Production
John Ferrell, U.S. Department of Energy

D

Recommendations from National Research Council Review of the FreedomCAR and Fuel Research Program, Phase 1

CHAPTER 2: MAJOR CROSSCUTTING ISSUES

Determining Priorities, Milestones, and Go/No-Go Decisions

Recommendation 2-1. An ongoing, integrated well-to-wheels assessment should be made of the Partnership's progress toward its overall objectives of reducing the nation's oil dependence and introducing hydrogen as a transportation fuel, if appropriate. This assessment should examine possible trade-offs between the individual goals of the fuel program and the vehicle program, as well as between short-term goals and long-term goals, and between energy sources, to guide future research priorities and, ultimately, national transportation energy policy.

Systems Analysis and Simulation

Recommendation 2-2. The FreedomCAR and Fuel Partnership should use its systems analysis capability routinely in the program management process, establishing goals, evaluating trade-offs, setting priorities, and making go/no-go decisions.

NOTE: Recommendations set entirely in bold are contained in the executive summary of the Phase 1 Report.

Recommendation 2-3. The FreedomCAR and Fuel Partnership should develop and refine its models for consumer behavior during a market transition to radically different vehicles and should also explore ways to enhance the effectiveness of its cost models.

Recommendation 2-4. The FreedomCAR and Fuel Partnership should assign responsibility for overall program management and for the complex analyses to support program management, such as technology assessments, goal checking, evaluating the broader impacts of the technologies on the major problems, commercialization assessment, and decisionmaking, among others.

Safety

Technical Teams

Recommendation 2-5. DOE should form a new crosscutting safety technical team with a mission that includes broad hydrogen-related safety issues not only for HFCIT but for the other DOE offices as well. The new team should incorporate the existing codes and standards technical team as a subteam. The other offices should assign a person to be responsible for safety and to interface with the safety technical team. The safety, codes and standards effort needs adequate resources so that it can accomplish the goals identified in its roadmap.

Vehicle Standards and NHTSA

Recommendation 2-6. NHTSA should begin its hydrogen R&D program in FY05 by focusing on the effects of hydrogen releases and other potential hazards with hydrogen-fueled vehicles as well as analyses and research to determine the right mix of system-level and component-level standards. NHTSA should also work with other U.S. and international safety groups to establish global standards for hydrogen-fueled vehicles.

Publication, Openness, and Safety Documents

Recommendation 2-7. DOE, USCAR, and NHTSA should prepare and maintain a bibliography of hydrogen-safety-related reports and papers and make that information available on their Web sites in a user-friendly manner. NHTSA and DOE should develop investigation protocols and have investigation teams ready to visit serious incidents anywhere.

Budget and Schedule

Recommendation 2-8. DOE should examine the budget and schedule estimates for each of the codes and standards deliverables and also for the other safety activities of the Safety, Codes and Standards program. To the extent that the budget and schedule are incompatible, changes should be reflected in the next update of the roadmap.

Learning Demonstrations

Recommendation 2-9. The FreedomCAR and Fuel Partnership should continue to develop prompt and effective channels of communication among its members to disseminate the learning from the demonstrations. The results should also be disseminated to supporting organizations outside the Partnership in order to promote widespread innovation and competition. But once the learning demonstration for a project has been carried out, the project should be reassessed to see whether further operation is warranted.

Recommendation 2-10. DOE management should keep the demonstration projects focused on their primary purpose—the accumulation, analysis, and dissemination of experience from the field. Safety should be stressed throughout the learning demonstration program, because an accident early on could attract publicity out of proportion to its true consequences.

Recommendation 2-11. Among the high priorities for feedback, DOE should identify precursor incidents that point to incipient safety problems and should develop appropriate methods for training first responders to deal with hydrogen-related emergencies.

Recommendation 2-12. The FreedomCAR and Fuel Partnership should develop effective channels of communication among its members to disseminate lessons learned and communicate to appropriate organizations outside the Partnership to promote in them a culture of innovation and competition within the developing support structure.

Goals and Targets

Recommendation 2-13. The program should perform high-level systems analyses that identify the potential, the challenges, and the specific research breakthroughs for alternatives that could achieve the program vision without requiring a hydrogen infrastructure, and it should use these results to help define R&D efforts and allocate funds within DOE.

Roles of the Federal Government and Industry

Recommendation 2-14. The FreedomCAR and Fuel Partnership and USCAR leadership should examine the effectiveness of the current process for transferring technology from DOE projects to within-the-industry activities and develop and implement procedures that will make such transfer as effective as possible.

FreedomCAR in the Policy Context

Recommendation 2-15. DOE should analyze the implications of alternative market interventions for the technical goals of the FreedomCAR and Fuel Partnership. These implications then could be included in DOE's policy deliberations.

Environmental Impacts

Recommendation 2-16. The DOE, in collaboration with the Environmental Protection Agency, should systematically identify and examine possible long-term ecological and environmental effects of the large-scale use and production of hydrogen from various energy sources.

CHAPTER 3: VEHICLE SUBSYSTEMS

Advanced Combustion Engines and Emission Controls

Recommendation 3-1. DOE should encourage the energy industry to become involved in establishing research parameters for the work on pure fuels that will be most relevant to real-world fuels expected in the marketplace.

Recommendation 3-2. DOE and the energy industry should develop refinery models for making tailored fuel blends.

Recommendation 3-3. Increased emphasis should be placed on novel emission control technologies, and the advanced combustion and emissions control technical team should plan for, analyze, and seek solutions for emission problems associated with emerging fuels, fuel infrastructure, and propulsion systems.

Fuel Cells

Recommendation 3-4. DOE should broaden its collaboration with industry, academia, and other government agencies on precompetitive, industry-wide technical issues and solutions. Stationary fuel cell developers should be included as well. For example, DOE could sponsor one or more conferences, workshops, debates, or forums to facilitate in-depth interactions or it could set aside some discretionary funds that would allow program managers to accelerate progress on promising new ideas.

Recommendation 3-5. To promote new fuel cell water and hardware imaging techniques that could address technical barriers, DOE should enhance its existing collaboration with the NIST Neutron Research Center. DOE should also determine whether similar capabilities exist at the national laboratories and related academic centers so it could capitalize on this significant analytical advancement.

Recommendation 3-6. DOE should expand activity and place a higher priority on membrane R&D, new catalyst systems, and electrode design (with the BES program). In particular, the national laboratories and other appropriate scientific centers should be focused on the fundamental failure mechanisms, including a better understanding of the chemistry, physics, and materials involved.

Hydrogen Storage

Recommendation 3-7. In view of the exploratory nature of the work and the need to take technical risk and thereby foster discovery, DOE should check progress at appropriate times with go/no-go decisions. In this way, new ideas are able to emerge and the most promising approaches are adequately supported.

Recommendation 3-8. The center of excellence research model should be carefully evaluated in parallel with peer review of the research. The committee believes centers of excellence are a good concept, but DOE should wait for an evaluation of the three centers' performance before expanding the concept to other areas of research.

Recommendation 3-9. In view of the risk posed to the entire hydrogen program by the need for a viable hydrogen storage system, the hydrogen storage technical team and the FreedomCAR and Fuel Partnership leadership team should report annually to all program participants, DOE, and Congress on the state of hydrogen storage technology worldwide relative to the goals and targets of the program.

Electrochemical Energy Storage for Electric Vehicles

Recommendation 3-10. DOE should direct more of its effort and funding for high-power batteries for HEVs to applied and long-term exploratory research rather than battery development.

Recommendation 3-11. A significantly larger effort and higher priority should be placed on searching for breakthrough technology in the area of high-energy batteries for electric vehicles.

Recommendation 3-12. In view of the potential benefits of a high-energy-density DLC in hybrid vehicles, the energy storage technical team, in conjunction with the electrical and electronics system technical team, should maintain an activity that explicitly monitors progress of international DLC research programs and should consider funding research in advanced DLC technologies.

Electrical Systems and Electronics

Recommendation 3-13. The EE technical team should play a leading role in coordinating the specifications for the interfaces among the many vehicle subsystems, using established standards where they exist and accelerating the development of new ones where they are needed.

Recommendation 3-14. The EE technical team should identify the R&D path leading to the motor cost goal, or it should reassess that goal.

Recommendation 3-15. The EE technical team should use its evaluation of the state of the art of HEV technology to update and establish the team's future research agenda and goals.

Recommendation 3-16. The EE technical team should develop a process for coordinating the diverse activities it is overseeing.

Recommendation 3-17. Integrating the electronics with the motor may well provide significant cost advantages. The EE technical team should consider these potential advantages and extend Table 3-6 to include aggressive targets for an integrated system in 2010 and 2015.

Recommendation 3-18. High-temperature power electronics and advanced thermal management systems will significantly impact the size, weight, cost, and reliability of the EE subsystems. FreedomCAR work in this area appears to be limited to the application of SiC devices to the Semikron inverter. The EE technical team should be aware of and leverage the work on high-temperature semiconductors, packaging, and thermal management being funded by government agencies at universities, commercial organizations, and the national laboratories.

Structural Materials

Recommendation 3-19. The only FreedomCAR effort on HSS should be careful monitoring of outside programs with the objective of adopting novel manufacturing and assembly methods to aluminum structures. This recommendation mirrors the previous NRC recommendation on the PNGV program.

Recommendation 3-20. The most important aspect of the stamped aluminum program is cost reduction, particularly for the feedstock material. Efforts in manufacturing should be limited until progress in the cost area has been achieved.

Recommendation 3-21. More extensive research programs on CFRPs, combined with the direct cooperation of the large fiber manufacturers, appears mandatory for any hope of success within the program time frame. Meanwhile, R&D for manufacturing of structures should continue.

Recommendation 3-22. Longer-term research programs in magnesium alloys should be funded because of the weight savings these materials could offer. Cast materials should be the primary emphasis, with limited exploratory work on wrought materials. Increased activity in this area is highly recommended.

Recommendation 3-23. The materials technical team should provide technical materials input to other technical teams—for example, electronics, the hydrogen on-board supply system, magnets, motors, fuel cell structural issues—where such input would be useful. The team has never been asked to do this, but it could be extremely useful to the overall program.

Recommendation 3-24. The materials technical team should provide models of weight reduction/cost trade-offs to the systems analysis and engineering team. This would help define the singular objectives for individual systems and allow some flexibility in the focus of cost reduction efforts.

Recommendation 3-25. Overall, since cost reduction is the main need in many of the materials programs, the committee suspects that research activities are of somewhat limited benefit. Thus, much of this research funding might better be expended on other more challenging research areas, such as hydrogen storage materials, batteries, fuel cells, and the infrastructure.

CHAPTER 4: HYDROGEN FUEL PRODUCTION AND DISTRIBUTION

Recommendation 4-1. The committee strongly recommends that the Hydrogen Technology R&D be fully funded at the $99 million level for the areas indicated in the FY06 Presidential budget request to Congress.

Recommendation 4-2. DOE should pay special attention to the transition from the current ICE fuels infrastructure to a nascent hydrogen economy. As part of this attention, the DOE should further focus the achievements of the fuel/vehicle pathway integration technical team by placing greater emphasis on the transition to hydrogen in its systems analysis work and should apply its systems capabilities to analyzing whether the cost goals for hydrogen production, established for a mature hydrogen economy, are appropriate for the transition. Specifically, this analysis should examine whether setting a hydrogen cost goal during the transition that is higher than the cost goal for a mature hydrogen economy would speed or impede the introduction of fuel-cell-powered vehicles.

Recommendation 4-3. The committee believes that significant development efforts should be directed to distributed hydrogen production, including natural gas reforming and electrolysis as well as exploratory work on other distributed generation options.

Recommendation 4-4. Even closer coordination with other DOE programs would be beneficial, including programs in the Office of Fossil Energy (FE) and the Office of Nuclear Energy, Science and Technology (NE). Representatives from FE and NE should be added to the fuel/vehicle pathway integration and hydrogen production technical teams, and FE and NE should be linked closely with systems analysis efforts in the Hydrogen, Fuel Cells and Infrastructure Technology program.

Recommendation 4-5. DOE should create a carbon capture and storage (CCS) system subteam (under the hydrogen production team) in the FreedomCAR and Fuel Partnership and make it part of the overall Hydrogen Fuel Initiative.

Recommendation 4-6. The goal of ±30 percent precision in estimating CO_2 capacity should be focused on geological storage.

Recommendation 4-7. DOE should strengthen the ties between managers of the CCS effort at HFCIT and managers at FE by developing a specific CCS program for hydrogen within FE. In addition, DOE should increase the shared management responsibility of the CCS program between EERE and FE.

Recommendation 4-8. The technical teams working on hydrogen production, delivery, dispensing, and storage should identify the unique R&D needs for hydrogen storage for production, as well as for delivery and dispensing, that are not being adequately addressed by the current project portfolio.

E

Acronyms

BES	(Office of) Basic Energy Sciences (DOE)
BOP	balance of plant
C&S	codes and standards
CAFE	corporate average fuel economy
CAR	Cooperative Automotive Research
CCS	carbon capture and sequestration
CFD	computational fluid dynamics
CFRP	carbon-fiber-reinforced polymer
CLEERS	crosscut lean exhaust emission reduction simulation
CNG	compressed natural gas
CO	carbon monoxide
CO_2	carbon dioxide
COE	center of excellence
CRC	Coordinating Research Council
DOE	U.S. Department of Energy
DOT	U.S. Department of Transportation
DPF	diesel particulate filter
DTI	Directed Technologies, Inc.
DTT	delivery technical team
E85	85 percent ethanol
EEA	Energy and Environmental Analysis, Inc.
EERE	(Office of) Energy Efficiency and Renewable Energy (DOE)

EPA	U.S. Environmental Protection Agency
EV	battery electric vehicle
FACE	fuels for advanced combustion engines
FC	fuel cell
FCFP	FreedomCAR and Fuel Partnership
FCHEV	fuel cell hybrid electric vehicle
FCVT	FreedomCAR and Vehicle Technologies (program)
FE	(Office of) Fossil Energy (DOE)
FFV	flexible fuel vehicle
FPITT	fuel pathway integration technical team
FY	fiscal year
GATE	Graduate Automotive Technology Education
GDL	gas diffusion layer
GFRP	glass-fiber-reinforced plastic
gge	gallons gasoline equivalent
GHG	greenhouse gas
GREET	Greenhouse Gas, Regulated Emissions, and Energy Use in Transportation (model)
GW	gigawatt
H or H_2	hydrogen
H2A	Hydrogen Technology Analysis (model)
HAMMER	Hazardous Materials Management & Emergency Response (facility)
HC	hydrocarbon
HEV	hybrid electric vehicle
HFCIT	Hydrogen, Fuel Cells and Infrastructure Technologies (program)
HFCV	hydrogen fuel cell vehicle
HFI	Hydrogen Fuels Initiative
HSS	high-strength steel
HyTrans	Hydrogen Transition (model)
ICC	International Codes Council
ICE	internal combustion engine
IEA	International Energy Agency
IGBT	insulated gate bipolar transistor
IGCC	integrated gasification combined cycle
kg	kilogram
kW	kilowatt
kWe	kilowatt (electric)

kWh	kilowatt-hour
Li ion	lithium ion
LPG	liquefied petroleum gas
LTC	low-temperature combustion
M85	85 percent methanol
MARKAL	Market Analysis (model)
MATT	Mobile advanced technology testbed
MEA	membrane electrode assembly
MOU	memorandum of understanding
MPa	megapascal
MSM	MacroSystem Model
MWe	megawatt (electric)
NAE	National Academy of Engineering
NAS	National Academy of Sciences
NE	Office of Nuclear Energy (DOE)
NEMS	National Energy Modeling System
NFPA	National Fire Protection Association
NGNP	Next Generation Nuclear Powerplant
NHTSA	National Highway Traffic Safety Administration
NIST	National Institute of Science and Technology
NO_x	nitrogen oxides
NPC	National Petroleum Council
NRC	National Research Council
NREL	National Renewable Energy Laboratory
OEM	original equipment manufacturer
ORNL	Oak Ridge National Laboratory
PBA	(Office of) Planning, Budget and Analysis (DOE)
PC	personal computer
PEIS	Programmatic Environmental Impact Statement
PEM	proton exchange membrane
PHEV	plug-in hybrid electric vehicle
PHMSA	Pipeline and Hazardous Materials Safety Administration
PM	particulate matter
PNGV	Partnership for a New Generation of Vehicles
PNNL	Pacific Northwest National Laboratory
PRD	pressure relief device
PSAT	Powertrain Systems Analysis Toolkit
PV	photovoltaic

R&D	research and development
RFP	request for proposal
RITA	Research and Innovative Technology Administration (DOT)
RSPA	Research and Special Projects Administration (DOT)
SAE	Society of Automotive Engineers
SBIR	Small Business Innovation Research
SC	(Office of) Science (DOE)
SCI	special crash investigation
SCR	selective catalytic reduction
SCS	safety, codes and standards
SER	strategic environmental review
SiC	silicon carbide
SMR	steam methane reforming
SNL	Sandia National Laboratories
SRI	Stanford Research Institute
STTR	small business technology transfer
SUV	sport utility vehicle
USABC	United States Advanced Battery Consortium
USCAR	U.S. Council for Automotive Research